AUSTRALIA
LAND BEYOND TIME

AUSTRALIA
LAND BEYOND TIME

Reg Morrison

Comstock Publishing Associates,
a division of Cornell University Press
Ithaca and London

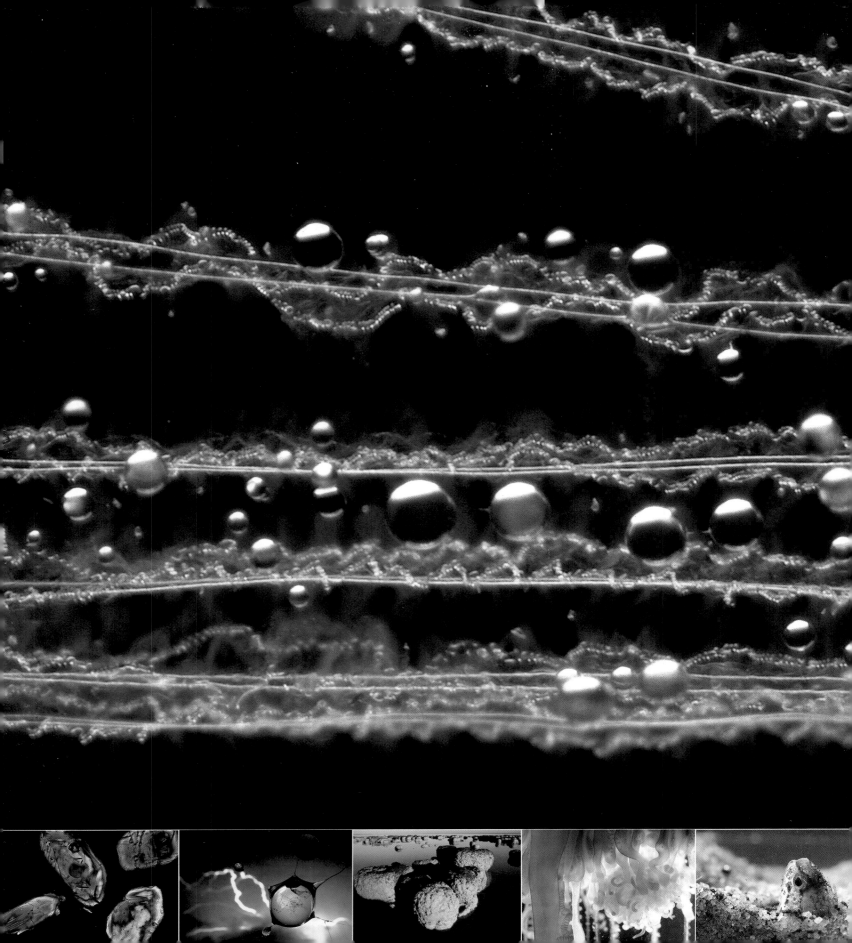

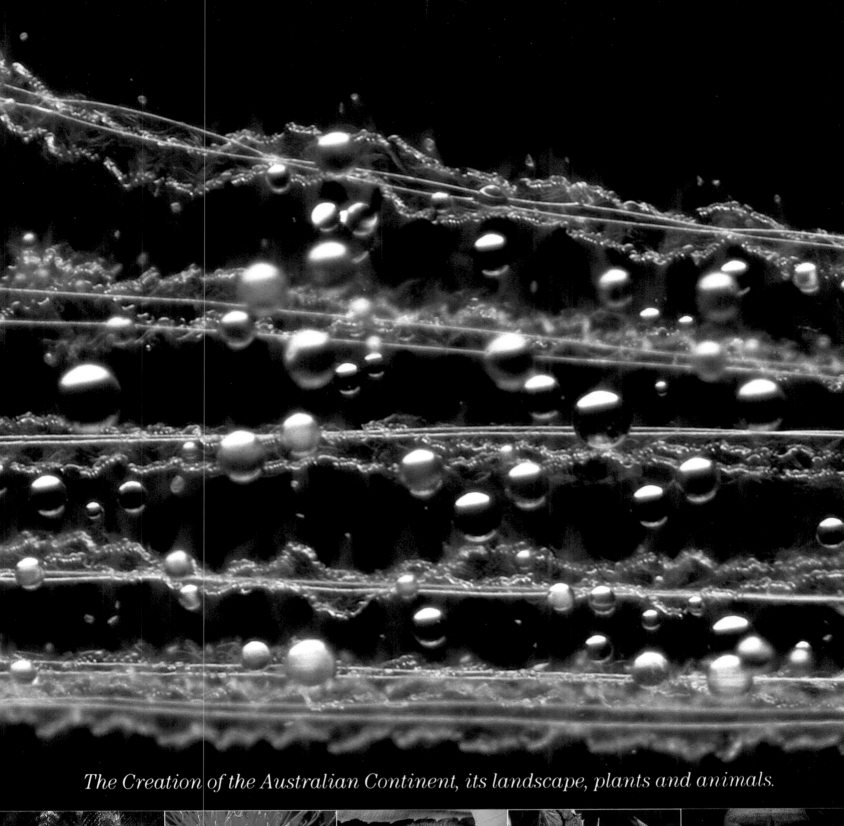

The Creation of the Australian Continent, its landscape, plants and animals.

CONTENTS

Earth's crust solidifies, oceans form and permanent landmasses appear; their historic remnants in Western Australia; cornerstones of a continent.

Life appears, and leaves its mark; Western Australia's historic fossils; their modern counterparts; biological evolution begins.

Life thrives, and pays a price; the Hamersleys, legacy of a global disaster; chlorophyll, the molecule that changed the world.

Landmasses accumulate; Australia's building blocks multiply, continental drift develops; evolution changes course.

Australia's foundations drift from pole to pole, weld with Antarctica; an evolutionary time bomb primed.

World climates plunge into disorder; ice engulfs Australia – in the tropics; 3,000 kilometre trail of evidence.

Evolution bounces back; multicelled animals crowd the fossil record – oxygen users arrive; lifeforms multiply.

Earth's continents pack together; Gondwana, the supercontinent of the south; Australia crushed; Ayers Rock and the Olgas.

GREEN PYTHON, *Chondropython viridis*, QLD.

Printed in Singapore

Library of Congress Cataloging-in-Publication Data
Morrrison, Reg.
    Australia: land beyond time/Reg Morrison.
      p. cm
    Includes bibliographical references and index.
    ISBN: 0-8014-8824-9
    1. Geology — Australia I. Title.
    QE340.M67 2002
    559.4 — dc21

Cornell University Press strives to use environmentally
responsible suppliers and materials to the fullest extent possible
in the publishing of its books. Such materials include vegetable-
based low-VOC inks and acid-free papers that are recycled,
totally chlorine-free, or partly composed of non-wood fibers.

Softcover printing 10 9 8 7 6 5 4 3 2 1

EVERLASTINGS, *Helipterum sp.*, W.A.

ACKNOWLEDGMENTS

A comprehensive diary of Australia's evolution has not
previously been attempted in this form and I am indebted to
a great many people for their assistance in interpreting and
simplifying technical material, and presenting it coherently
in this condensed version.

I would like to express my deepest gratitude to
DR. ALEX RITCHIE , Divisional Head of
Earth Sciences, Australian Museum, Sydney, and
DR. MARY WHITE for their generous advice
and unstinting efforts to save me from the
minefield of technical errors that confront
any writer in this field.

Others I would specially like to thank for their
assistance with the text are DR. MICHAEL ARCHER
(University of New South Wales),
DR. CLIVE BURRETT (University of Tasmania),
PROF. LARRY FRAKES (Adelaide University),
DR. LYAL HARRIS (University of Western Australia),
DR. DON MACPHEE (La Trobe University),
DR. ALAN THORNE (Australian National University),
DR. TONY THULBORNE (University of Queensland),
DR. MALCOLM WALTER (Macquarie University) and
MR. KINGSLEY DIXON (Kings Park Botanical Gardens, W.A.)

Planetary Formation:
DR. A.J.R. PRENTICE, Monash University, Melbourne.
PROF. A.E. RINGWOOD, Aust. National University, Canberra.

Geology and Geochemistry:
DR. BILL COMPSTON, Aust. National University, Canberra.
DR. LYAL HARRIS, University of W.A., Perth.
MR. PETER KINNY, Aust. National University, Canberra.
MR. KEN PLUMB, Bureau of Mineral Resources, Canberra.
DR. LIN SUTHERLAND, Australian Museum, Sydney.
DR. IAN WILLIAMS, Geological Survey of W.A., Perth.

Palaeomagnetism and Plate Tectonics:
DR. CLIVE BURRETT, University of Tasmania, Hobart.
MR. DAVID CLARK, Mineral Physics, C.S.I.R.O., Sydney.
DR. BRIAN EMBLETON, Mineral Physics, C.S.I.R.O., Sydney.
DR. JOHN GIDDINGS, Bureau of Mineral Resources, Canberra.
DR. CHRIS KLOOTWIJK, Bureau of Mineral Resources,
Canberra.
DR. PHIL SCHMIDT, Mineral Physics, C.S.I.R.O., Sydney.
PROF. JOHN VEEVERS, Macquarie University, Sydney.

Palaeoclimate (Ice Ages):
PROF. LARRY FRAKES, Adelaide University,

Proterozoic Biology and Stromatolites:
DR. ROGER BUICK, University of W.A., Perth.
MR. JOHN DUNLOP, geological consultant, Perth.
DR. BRIAN LOGAN, University of W.A., Perth.
MS. LINDA MOORE, University of W.A., Perth.
DR. MARJORIE MUIR, C.R.A. Pty Ltd, Canberra.
DR. PHILLIP PLAYFORD, Geological Survey of W.A., Perth.

Genetics:
DR. DON MACPHEE, La Trobe University, Melbourne.

Plant Evolution:
DR. MARY E. WHITE, Sydney

Animal Evolution:
DR. MIKE ARCHER, University of N.S.W., Sydney.
DR. ALEX RITCHIE, Australian Museum, Sydney.
PROF. BRUCE RUNNEGAR, University of New England, N.S.W.

Invertebrates
DR. MIKE GRAY, Australian Museum, Sydney.
DR LOUIS KOCHS, West Australian Museum, Perth.
DR. DAVID MCALPINE, Australian Museum, Sydney.
DR. KEN McNAMARA, W.A. Museum, Perth.
DR. ROBERT RAVEN, Queensland Museum, Brisbane.

Fish:
DR. JOHN PAXTON, Australian Museum, Sydney.

Dinosaurs:
DR. RALPH MOLNAR, Queensland Museum, Brisbane.
DR. TOM RICH, National Museum of Victoria, Melbourne.
DR. TONY THULBORNE, Queensland University, Brisbane.
DR. MARY WADE, Queensland Museum, Brisbane.

Reptiles:
DR. HAL COGGER, Australian Museum, Sydney.

Birds:
DR. RICHARD SCHODDE, C.S.I.R.O., Canberra.

Mammals:
DR. MIKE ARCHER, University of N.S.W., Sydney.
DR. ROD WELLS, Flinders University, Adelaide.

Flora:
DR. ALISON BAIRD, Perth.
MR. ANDREW BROWN, C.A.L.M., Perth.
MR. KINGSLEY DIXON, King's Park Botanic Gardens, Perth.
MR. ROBERT DIXON, King's Botanic Gardens, Perth.
MR. CLYDE DUNLOP, N.T. Conservation Commission, Darwin.
DR. NEVILLE MARCHANT, W.A. Herbarium, Perth.

SPECIAL SUBJECTS

Salamander Fish:
DR. BRAD PUSEY, University of W.A., Perth.

Queensland Lungfish:
DR. ANNE KEMP, Queensland Museum, Brisbane.

Gastric Brooding Frog:
MR. MICHAEL TYLER, Adelaide University,

Honey Possum:
DR. RON WOOLLER, Murdoch University, Perth.

Numbat:
DR. TONY FRIEND, C.A.L.M., Perth.

Human Prehistory:
DR. ALAN THORNE, Aust. National university, Canberra.

I would like to add a special note of thanks to
rangers and other officers of the various parks
and wildlife services both State and Federal
who helped my wife and I during our travels,
and to the staffs of various State herbariums
and forestry offices who went out of their way
to help us with plant identifications.
We are grateful also to the owners and managers
of the following pastoral leases for their
hospitality and for giving us access to
sites that were crucial to this story:

Peter and Valmai Kopke, *Carbla Station, W.A.*
Sandy and Carol McTaggart, *Mt. Narryer Station, W.A.*
David and Maureen Halleen, *Boolardy Station, W.A.*
Peter and Jean Burton, *Meeberrie Station, W.A.*
Fred Quartermaine, *Mt. Edgar Station, W.A.*
Ken and Vicki Hasted, *Christmas Creek Station, W.A.*
Brian Fielder, *Brooking Springs Station, W.A.*
Sean and Toni Coutts, *McArthur R. Station, N.T.*
Mick and Ann Seymour, Lloyd Campbell and Graham
Stabler, *Riversleigh Station, Qld.*
Don Pinwill, *Yaramulla Station, Qld.*
James and Marjorie Lord, *May Downs Station, Qld.* and
also Herman and Mavis Malbunka for their hospitality and
assistance during our stay at *Tnorula (Gosse Bluff), N.T.*

My thanks also to Joe and Diane Vavryn of *The Caves,
Queensland,* for their hospitality and help in
photographing the bats and pythons of Mount Etna.
Finally we would like to thank old friends,
Denis O'Meara for his hospitality and assistance
in Perth and the Pilbara, Phil Mathews for
his editorial expertise and support throughout
this four-year project, and Maureen and Cec Murphy
for their assistance in preparing the text.

I am grateful to the following for the use
of their excellent photographs:

Jean-Paul Ferrero, *Auscape International (p. 237).*
John Fields, *Australian Museum (p. 180).*
Jim Frazier, *Mantis Films (p. 138, p. 145).*
Mike Gillam, *Auscape International (p. 261).*
Dr. Rob Morrison, *S.A. College of Advanced Education
(p. 298–9).*
Mike Tyler, *Adelaide University (p. 129),
and to Peter Schouten for his superb pencil
drawings of dinosaurs and giant marsupials.*

*All photographs were taken with Nikon
cameras fitted with Nikkor lenses ranging
from 16 mm to 800 mm.
Kodachrome film was used exclusively.
I am grateful to Nikon's Sydney agents,
Maxwell Optical Industries, and to
Mr. Peter Saidey in particular,
for their assistance with equipment.*

A Traveller's Guide...

GANTHEAUME POINT, BROOME, W.A.

A CAUTIONARY WORD to intending travellers: this journey frequently edges into relatively unexplored territory and several hypotheses stray considerably from customary pathways. For this reason it may not always be a comfortable trip for the pedant or the faint of heart, but it is surely worth the odd stumble. The basic information here, I hasten to add, came from specialists of the highest calibre, scientists and academics who gave unstintingly of their time and energy to ensure that their information was accurately conveyed. In many cases, however, conditional riders and other qualifiers that they properly attached to their extrapolations have been omitted here. Put another way, I have deliberately removed some of the safety nets of scholarship so that the view is not obscured for the average reader.

In reality there is now little doubt that the events upon which this book hinges did occur. Unresolved in some instances are questions regarding the precise timing of events, their duration and their exact magnitude. But in view of the massive volume of sound information that is now available, it would be sad indeed to miss the majesty of the overview that it offers, merely for the sake of academic consensus on the details.

One major stumbling block to an appreciation of the scale of this journey is our limited perception of time. Each of us looks so briefly at the world from within the constraints of a single lifetime. While we readily subdivide our time into smaller units such as years, days, minutes or even milliseconds, it is much harder for us to see beyond the span of our own existence.

But the history of the Earth is so vast that human time scales become meaningless: even a mere million years is difficult to grasp.

To maintain some sense of this larger scale and to translate it into measures that we can more readily comprehend, the 4.6 billion (4,600 million) year lifetime of the planet is scaled down to a single day. The passage of aeons is ticked off against a 24-hour clock.

Australia's geological record sprawls across 90 per cent of this time scale. Not only is it the longest geological record in the world, but a startling fact has emerged: the oldest bits of Australia have drifted at least 100,000 kilometres – more than twice round the globe – in that time.

The development of accurate dating techniques and the ability to read original magnetisation patterns that remain fossilised within some rocks have made it possible to plot both the time and latitude at which certain geological events occurred. By joining together a number of these plots – provided they are chronologically close – a path may be traced against a time scale to show the variation in a continent's distance from the poles: in other words, the north–south component of its continental drift.

Such evidence of the massive wandering of continents over the face of the planet has suddenly begun to push formerly unrelated pieces of the geological jig-saw together to make a coherent picture for the first time.

In the case of Australia, this continental wander path is only one of a number of possible interpretations of the paleomagnetic evidence, but although there is often doubt about the polarity of the data, there is little doubt about the range of latitudinal drift and the huge distances Australia has journeyed. Australia's voyage, traced out by the grey line in the graph below, has been an epic one indeed.

AUSTRALIA'S WANDER-PATH
4.6 billion years : 24 hours

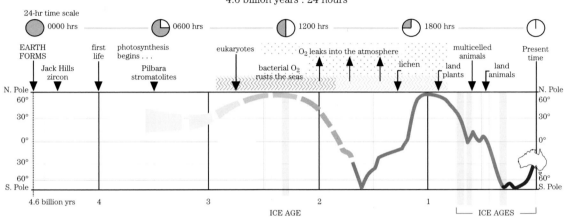

FROZEN MAGNETISM

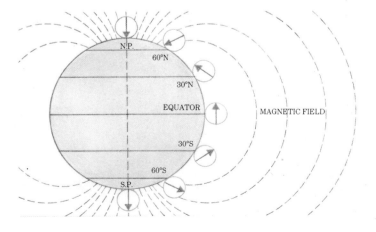

The Earth's magnetic field is shaped rather like an apple, curving inward towards the core at both magnetic poles. At these, a compass needle that is free to pivot in a vertical plane will point straight up, or down, according to its polarity: at the equator the same needle will remain horizontal. Between these two extremes however, it will adopt an attitude that corresponds to the field angle at that latitude.

Molecules of iron in molten rock such as volcanic lava also align themselves with the local magnetic field as they link together during crystallisation. In this way they record the latitude at which the rock solidified. By correlating ages and latitudes for a sequence of magnetised rocks it becomes possible to chart the probable drift pattern of a continent.

The Earth's magnetic field is probably caused by powerful electric currents generated in the molten iron and nickel of its outer core by the daily rotation of the mantle and crust.

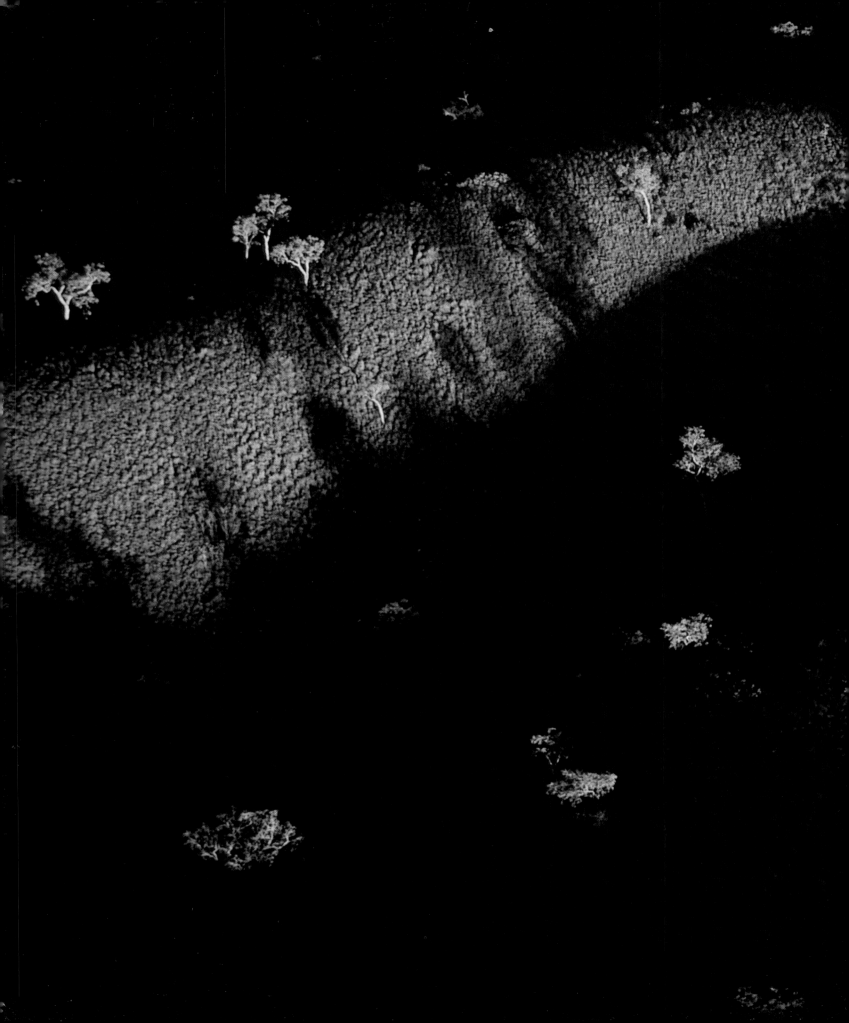

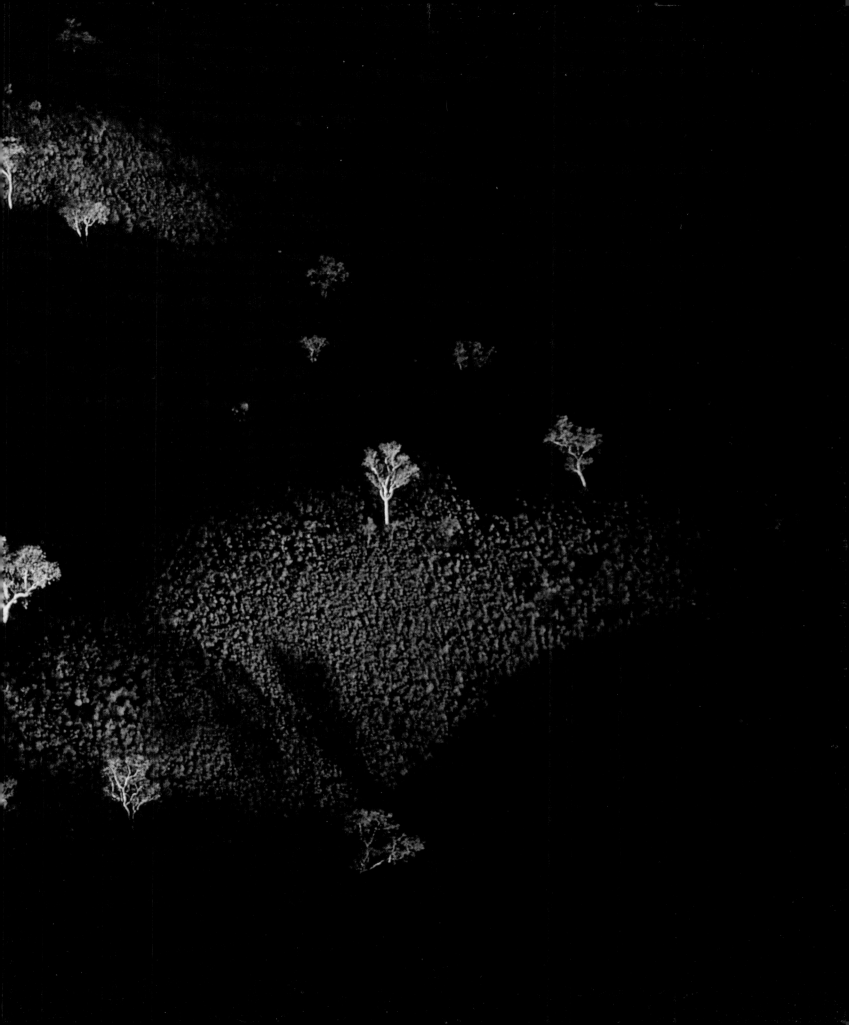

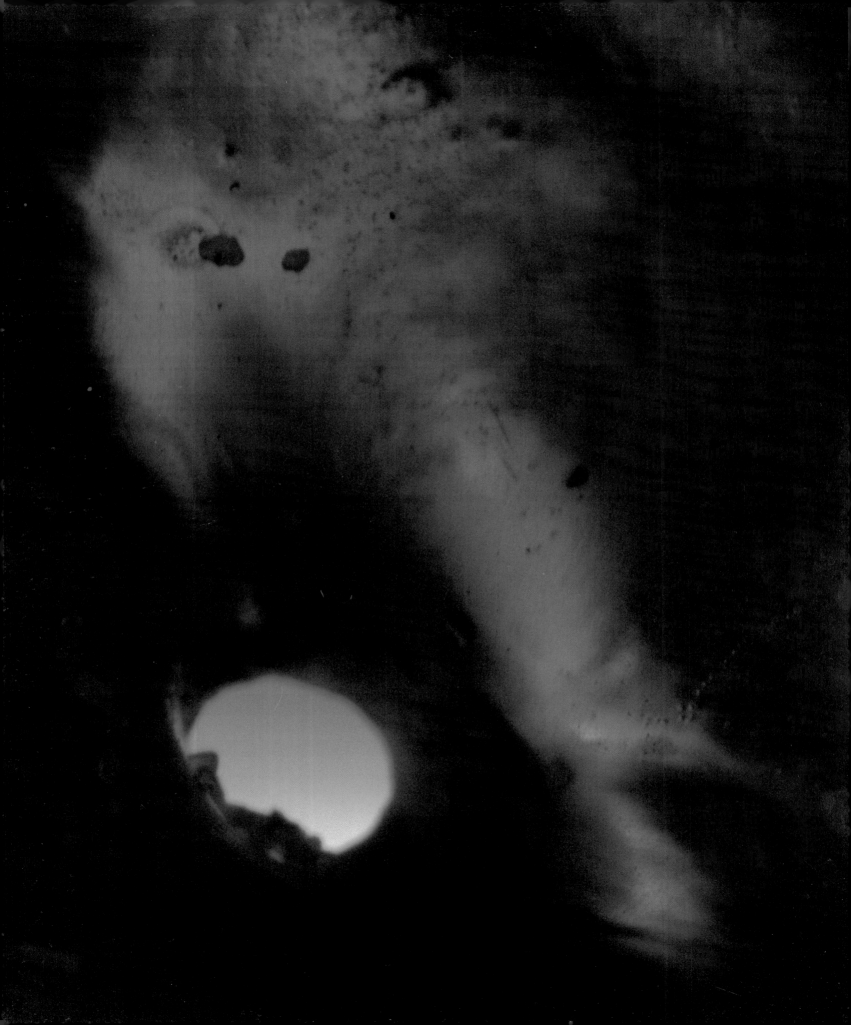

SIMPSON DESERT SAND, S.A.

The lining of an abalone shell conjures with the light from the setting sun to re-create, in microcosm, the scene from which we trace our origins. The Earth coalesced in a setting like this from clouds of gas and icy star 'dust' shed by the new-born sun about 4.6 billion (4,600 million) years ago. The main constituents of the shell – calcium, carbon and oxygen – were abundant in the original solar material, and in this combination (calcium carbonate) became a characteristic by-product of primordial life.

Another crucial constituent of the solar 'dust' was silicon. Combined with oxygen it provided the buoyancy that made continents a permanent feature of the planet's surface. It is our geological home. Lit by polarised light, grains of desert sand (ABOVE) glow with colour because of their silica content.

PROLOGUE

THIS IS THE STORY OF AUSTRALIA. It begins so long ago that it also provides us with what amounts to a diary of the entire planet – a diary written on the stony fabric of the continent itself, and into the antique web of life that now inhabits it.

Our story begins about 4.6 billion (4,600 million) years ago in the icy void of deep space, near the edge of a spiralling cluster of stars. One of its older giants has exploded, and the shock wave has begun to surge through the dusty remains of a previous explosion. Compression stirs the dust cloud into eddies. One of these will form our solar system.

The body of slowly rotating gas, dust and ices begins to contract and accelerate, like a pirouetting skater who gradually folds her arms against her body. During the next 100 million years this mass will shrink to one ten thousandth of its original size, leaving a broad skirt of dusty material drifting slowly round its equatorial waist.

Meanwhile, compression ignites the hydrogen within the core, and a molecule-building process known as atomic fusion begins, heating the core until it glows. In the growing light from this, our Sun, we see a number of satellites of various sizes embedded in the skirt of rotating debris. Third from the Sun is our planet, Earth.

Bombarded by vast quantities of solar debris lying in its orbital path, and tortured from within by the growing force of its gravity, the young planet soon becomes heated to melting point, and its primary constituents begin to rearrange themselves. Heavy elements such as nickel, iron and manganese sink towards the core, while lighter ones, such as silica and aluminium, rise to the surface. Meanwhile, billowing from a multitude of surface cracks and vents, prodigious volumes of steam and other gases begin to form Earth's primitive atmosphere.

The Earth's gravity is still too weak to retain all of these gases, and some of the lighter ones, including water, dissipate into space. The seas that will ultimately cover 70 per cent of this planet, setting it apart from all others, will be largely delivered by comets.

Around 4 billion years ago, Earth approaches its final size. The bombardment eases, the crust cools, and surface water accumulates…

4.6 – 3.5 billion years

THE STORY OF AUSTRALIA *begins early in the life of the planet. If the 4.6-billion-year history of the Earth is matched against a 24-hour time scale, then Australia's continental seeds are sown in the small hours of the morning...*
At 1.00 a.m. on this time scale the planet is still in a process of birth. Meteorites bombard it incessantly, and heat from these impacts, and from atomic decay in the planet's pressurised core, fuses it into a ball of iron-rich magma. From a distance it looks like an old-fashioned pudding just out of the cooking cloth. Steam pours from the crust as gases are driven from the fractionating interior, and clouds soon envelop the planet. Lightning begins to crackle through the turbulent atmosphere and thunder becomes incessant. When at last the surface cools sufficiently for rain to fall, a torrent begins. The cometary bombardment eases around 2.30 a.m. on our 24-hour stopwatch. Earth's crust cools and water begins to collect in the multitude of craters that scar its face. Massively augmented by the continued fall of icy solar debris, these crater lakes soon link into warm, soupy seas. Rivers begin to carve into the land and carry the sediment away.
Australia's story begins with several grains of those primordial river sands.

These are the oldest known fragments of the Earth's crust. They are minute crystals of zircon that began to solidify more than 4.1 billion (4,100 million) years ago near what is now the western edge of Australia.

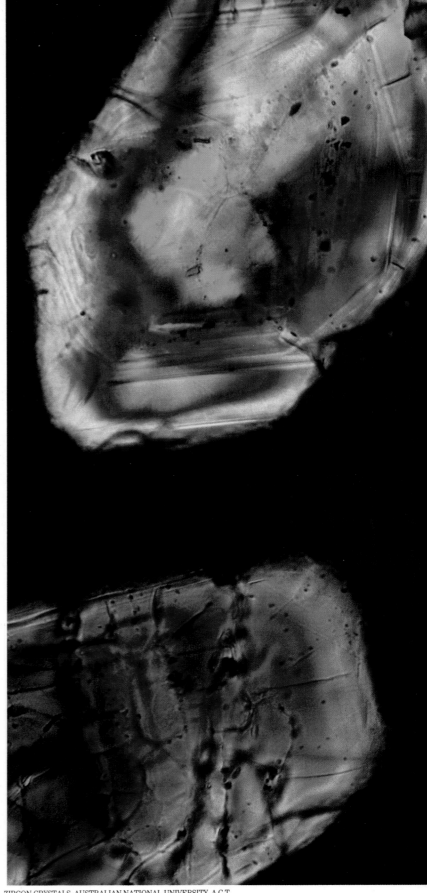

ZIRCON CRYSTALS, AUSTRALIAN NATIONAL UNIVERSITY, A.C.T.

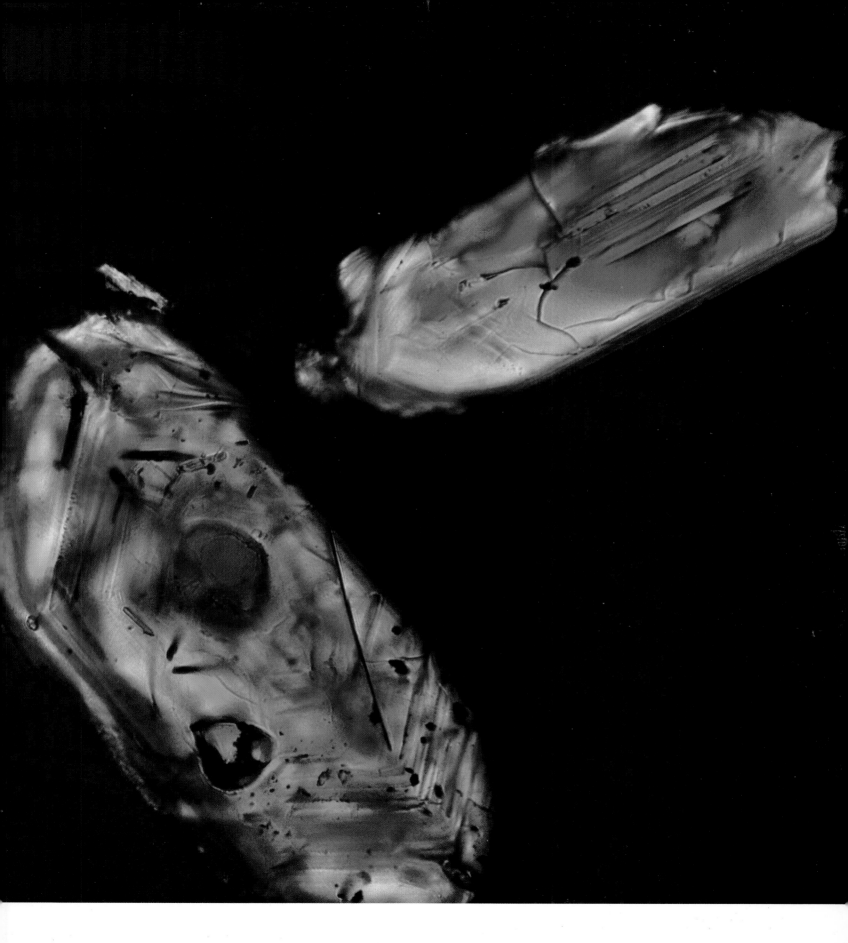

SEEDS

ZIRCON CRYSTAL, A.N.U., A.C.T.

SEEDS

GATHERING DUST IN A UNIVERSITY filing cabinet in Australia's Federal capital, Canberra, are several resin discs and microscope slides of extraordinary significance. Each disc and slide glitters slightly owing to a fine dusting of zircon crystals embedded in its surface. Many of these crystals are more than 4 billion years old, and one has recently been dated at more than 4.4 billion years old. As a tiny fragment of Earth's original crust, this one represents geology's Holy Grail.

All of these microscopic samples came from sites in the Narryer Range and at Jack Hills near the western coastline, where a slab of ancient land crust protrudes through a shroud of younger sediments. So this is where the story of Australia rightfully begins.

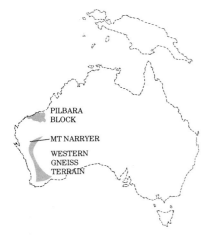

PILBARA
BLOCK

MT NARRYER

WESTERN
GNEISS
TERRAIN

Dawn silhouettes an outcrop of the Earth's oldest land crust at Mount Narryer (LEFT), in Western Australia, beneath trails inscribed on the film by the south polar starfield during a ten-hour time exposure. The sediments that form Mount Narryer's rocks were laid down by a river under northern stars more than 3.6 billion (3,600 million) years ago, in an atmosphere that would have been totally poisonous to us.

(ABOVE RIGHT): The core of this tiny zircon crystal found at Mount Narryer began to form more than 4.16 billion years ago. Lit by polarised light the zircon glows with rainbow fire, betraying its silicate nature. The dark hole near the centre of the crystal is the scar left by the dating process.

Wedged among the boulders on the ridge line that runs north from Mount Narryer, I watched the autumn stars as they began their nightly rotation about the craggy summit. Ahead of me was a journey in time that would retrace the progress of Australia, from the zircon relics of the Earth's primitive crust to its modern continental form.

Only recently has such a journey become possible, partly because of the development of dating techniques capable of accurately ticking off time scales so enormous as to be far beyond an easy grasp by the human mind. Zircons, for example, may now be dated through tell-tale traces of uranium locked within the onion-skin layers of each crystal during its growth. Uranium's giant atoms undergo constant, measurable decay, owing to their nuclear instability. Shedding sub-atomic particles with the regularity of a ticking clock, the uranium changes gradually from one form (isotope) into another until it eventually becomes lead. A crystallisation date can be deduced by measuring the ratios of the various isotopes within each crystal layer. A similar clock mechanism is sometimes offered by one or two companion elements so that dates may be double checked.

SITE T, MOUNT NARRYER, W.A.

FOUNDATIONS OF MOUNT NARRYER, W.A.

(RIGHT): The layered rocks that form the summit of Mount Narryer were deposited as river sediments between 3.5 billion and 3.6 billion years ago and include small boulders which, in cross section, show fractured layers that formed earlier mountains.

(ABOVE LEFT): Codes painted on the rocks pinpoint sites from which samples were taken during the dating programme that identified Mount Narryer as part of the oldest land crust in the world.

(LEFT): The granitic rocks that surface in layers along the north-eastern flanks of the Narryer Range are part of the silica-rich platform on which the Mount Narryer sediments were laid. They helped to push the region above sea level some 3.65 billion years ago.

Judged by this dating process, the oldest of the Narryer and Jack Hills zircons began to form at a time when the Earth's crust was under constant assault by solar debris from without, and by increasing volcanism from within. Careful isotopic analysis of the oxygen atoms present in the oldest zircon suggests that even at this time there was liquid water present on the Earth's surface. Similar analysis of its silica component suggests that it may well have been born in the granitic foundations of a shred of primordial land crust.

The Mount Narryer sediments appear to have been laid as river-bed sands on top of just such a granitic foundation. However, this solidified much later, around 3.65 billion years ago, and has since been severely deformed. Its crumbling layers now protrude along the north-eastern base of the range like a graveyard of tilted tombstones.

SULPHUR SPRINGS, PILBARA, W.A.

The sulphates that encrust the rocky bed of Sulphur Springs (ABOVE) in Western Australia's Pilbara not only betray the mineral wealth of the region, they are now believed to embody the biological wastes of some of the planet's oldest life forms. Bacterial fermenters seem to have thrived in most sulphurous volcanic vents and craters at that time. Graphic evidence of that volcanism shows in a nearby creekbed (RIGHT). The rounded shapes are the remains of giant 'pillows' of lava that froze on contact with sea water during a submarine eruption some 3.5 billion years ago. The region was then in the process of emerging from the sea to become one of the planet's oldest land masses.

PILLOW LAVA, PILBARA, W.A.

SEEDS

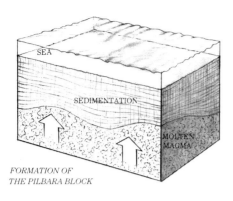

SEA

SEDIMENTATION

MOLTEN MAGMA

*FORMATION OF
THE PILBARA BLOCK*

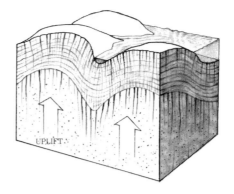

UPLIFT

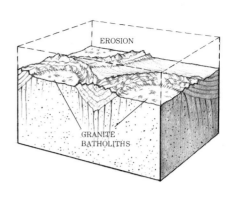

EROSION

GRANITE BATHOLITHS

GRANITE BATHOLITH, PILBARA, W.A.

The ancient Pilbara seabeds were finally lifted out of the water by an accumulation of buoyant, silica-rich granites in the crust beneath them. Scraps of these old seabeds still remain (BELOW), draped about the edges of the huge granitic platforms that pushed them to the surface. Most of the granite platforms themselves have also been laid bare by erosion and form the huge, boulder strewn plains of the northern Pilbara (LEFT).

No source for the river and its sediments has yet been established, nor has the primary rock from which the zircons were eroded been found. But the search continues. An outcrop of similar age has been discovered 80 kilometres to the north-east at Jack Hills, and another, almost as old, near Greenbushes in the south-west corner of Western Australia. All three sites are linked by a long, narrow shred of slightly younger material that forms a toughened edge on the continental 'shield'. Known as the Western Gneiss Terrain, this shred dips to the east beneath younger rocks, while to the west it halts abruptly at a huge fracture line which is parallel to the coast for a thousand kilometres. Clearly 'Australia' once extended further west before this extension was detached by cataclysmic shearing. Mounting evidence suggests that the missing portion now lies crushed and buried beneath the mountain wilderness of eastern Tibet, nearly 8,000 kilometres away. Such evidence demonstrates what gigantic forces have helped to shape the face of the planet.

Buoyant, and riding as permanent rafts on conveyor belts of recycling oceanic crust, older fragments of land crust, such as those in Western Australia, have wandered tens of thousands of kilometres over the face of the planet. Many other modern landmasses, notably Greenland, Canada, Africa, India and Siberia, have accumulated about such ancient cores. Pushed this way and that by convection currents in the molten mantle, and jostled by adjoining crustal plates, all continents have been repeatedly squeezed, fragmented and rearranged.

For the emerging landmass represented at Mount Narryer it has been an epic journey to continental maturity. Australia's mass has many times been perforated by huge volcanic boils, crumpled into alpine mountain chains, scourged by glaciers, inundated, and scorched to desert. As continental histories go, it is perhaps no more spectacular than that of other continents, but as a heritage it has been better preserved because of Australia's stability. It has been more visibly preserved too, because of the present lack of ice, water and vegetation.

PILBARA, W.A.

CHERT WALLS, PILBARA, W.A.

Long cracks appeared in the Pilbara seabed as the granite platforms rose, and the highly mineralised seawater of those times then slowly clogged the cracks with its soluble silica. These silica-filled cracks still pattern some Pilbara hillsides, though now in relief. Erosion has worn away the seabed sediments leaving the silica protruding in walls of coloured chert (ABOVE).

The ridge of gaudy, iron-stained silica known as the Marble Bar (LEFT), near the town of the same name, is one of the best known landmarks of the North-West. Originating as layered sediments on the sea floor, it was lifted above sea level and tilted vertically by the rise of the big granite platform that now flanks it to the east.

No clues remain of Australia's birthplace on the face of the young planet but the picture of it half a billion years later is slightly clearer. We know that the climate was then relatively warm and that a new land-mass was beginning to form in the seas nearby. Huge molten plumes of silica-rich granite, known as batholiths, had pushed up beneath the sea-bed. This now forms the basement of the region of north-western Australia known as the Pilbara. Erosion has rubbed away the covering from most of the old granite domes, but draped about them, in the 'crevices' between, are the crumpled remnants of the original sea beds. They form the disordered hills that now characterise much of the northern Pilbara. Great seams of silica, originally deposited in the cracks that opened as the seabed rose, now subdivide the hillsides with walls of chert.

Of far greater significance for us, however, is a tiny set of blemishes that became fossilised at the edge of one of these seabeds some 30 kilometres to the west ...

THE MARBLE BAR, COONGAN RIVER, W.A.

ALMOST A QUARTER *of our 24-hour time*
scale has elapsed since the solar system was
born. The sun's glow reveals its brood of orbiting
satellites. On Earth it is a gloomy dawn, though
the gigantic electrical storms of earlier times
have subsided and a little sunlight occasionally
filters through. At night, breaks in the cloud
reveal glimpses of the moon, the Earth's sole
satellite. They are very close together and the
moon's gravitational pull distorts both the
Earth's crust and its film of surface water,
making them slightly egg-shaped. Though the
Earth is rotating, the disortion remains
stationary in relation to the moon. This forms
two tidal surges, one on each side of the
spinning Earth.

Meanwhile, in steamy, sulphurous pools, a
new chemistry has appeared. Its molecular
origins are embedded in the cosmos and hinge
upon the unique ability of the carbon atom to
form flexible bonds with other atoms, linking
them into long molecular chains. Even more
important, some of these molecular chains can
replicate themselves.

Just before 6.00 a.m., in the weak light of the
young sun, this chemistry leaves its faint
signature on the glistening silt of a shallow,
tidal lagoon.

Many of Earth's first amino acids, the building blocks of
life, were probably hammered together by lightning in
the awesome electrical storms that constantly roamed
the steamy, primitive atmosphere. The spider web on
which these droplets hang is pure amino acid.

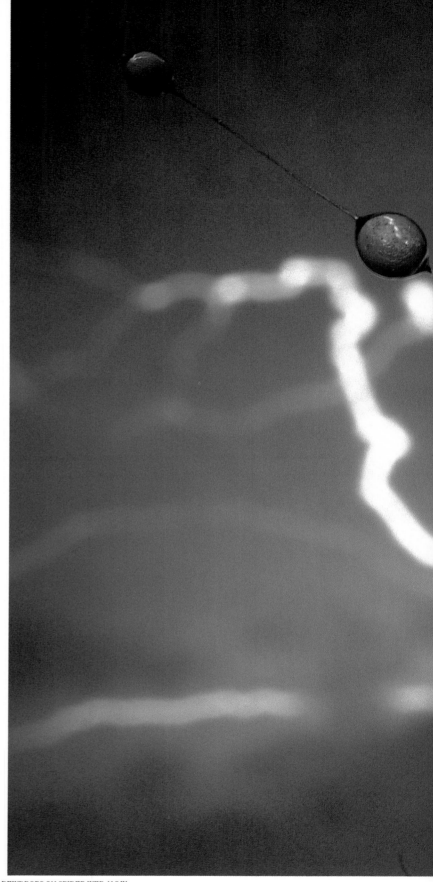

DEWDROPS ON SPIDER WEB, N.S.W.

SIGNATURE

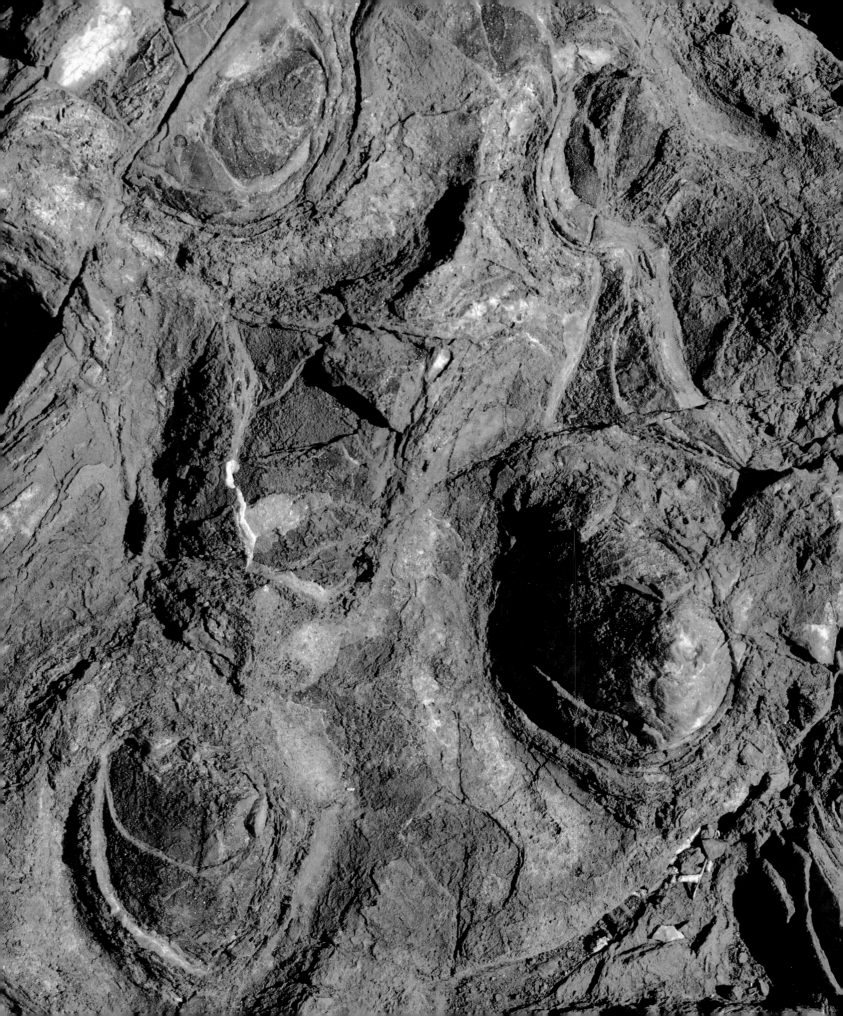

FOSSIL STROMATOLITE, NORTH POLE, W.A.

SIGNATURE

THE PATH IS STEEP and climbs between puffballs of spinifex and the bulbous outcrop of a submarine lava flow. It ends on a ridge at a small, rust-coloured outcrop of layered sediments. These are the 3.5 billion-year-old remains of a silt-filled tidal bay that was once ringed with smoking volcanoes and brooding lava hills. Sealed in by later lava flows, this muddy lagoon became a time capsule of unique significance. It has remained almost untouched by the crustal forces that usually disfigure or obliterate such relics.

Faithfully preserved within the thin layers of these sediments is a curious array of small irregularities which originally would have patterned the muddy floor of the bay. They represent a tiny hiccup in an otherwise peaceful period of sedimentation. At one place, beside a miniature fault line, a complete set of these irregularities has cracked gently away from the rest of the bedding. It forms a tiny domed column. Sliced vertically by recent weathering it reveals its layered structure, each successive layer building upon the hump in the layer beneath it. Nearby, in the same rock layer, there are other domes and columns. Amid the primordial geology of north-western Australia it is a strange form indeed.

It is remote, hot and arid there, and it is best known as a mining centre. On the few maps that show such detail it bears the wry title, North Pole. The explanation of the North Pole's enigmatic domes and columns lies some 800 kilometres to the south-west where, in a warm, turquoise recess of Shark Bay, similar structures occur in abundance. Though many times the size of the Pilbara one, these domes and columns, when cut lengthwise, show similar layering. They are the modern rubbish dumps of one of the world's most ancient life forms, microscopic organisms known as cyanobacteria. Shark Bay's cyanobacteria are direct descendants of the builders of the tiny structures that now show on the North Pole ridge line.

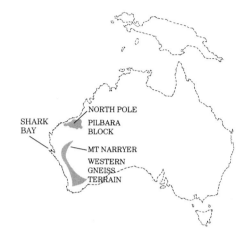

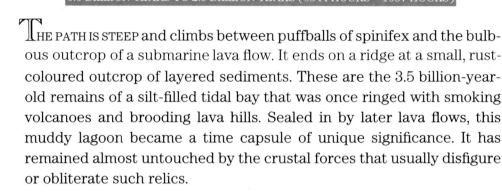

The earliest tangible proof of life on Earth consists of a few small bumps in a layer of siltstone in north-western Australia (LEFT). They are fossilised stromatolites, originally the muddy accretions left by small clusters of bacteria that lived along what was a shoreline some 3.5 billion years ago. One such lump, neatly sliced in half by weathering has gently cracked away from the cliff face to display its most basic form, a layered column (ABOVE RIGHT). These historic finds were made in an area known as the North Pole — a wry allusion to its isolation and summer heat.

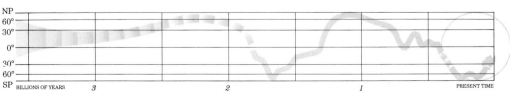

SIGNATURE

The growing light of the young Sun was probably the key to the first appearance of stromatolites like these. Their builders lived in shallow water and almost certainly depended on the energy in sunlight to drive their chemical processes. The only major colony of their descendants lives on the west coast of Australia (OVERLEAF).

MOUND BUILDERS

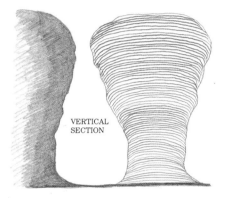

VERTICAL SECTION

NOON

MIDNIGHT

DAWN

The bacteria that build stromatolites depend on the energy in sunlight to power their feeding and growth. During the night these processes cease, allowing tide-borne silt to gather on the surface of the dormant colony.

At dawn the bacterial cells, which are commonly elongated and hair-like, find their way through the silt grains and re-establish the colony on the upper surface once more. Their calcium waste, secreted as a carbonate, is left behind during this movement and hardens into a cement. Layers of silt trapped in this way accumulate daily, growing into mounds and ultimately into columns.

Growth rates measured in Shark Bay average less than 0.5 mm a year, which suggests metre-high columns should be more than 2,000 years old.

FOSSILISED STROMATOLITES, NORTH POLE, W.A.

STROMATOLITES, SHARK BAY, W.A.

Living much as their ancestors did 3.5 billion years ago, cyano-bacteria feed on nutrients in the highly mineralised water of their habitat and secrete small quantities of carbonate waste. This glutinous waste traps particles of tide-borne silt, which build up layer by layer beneath the filmy mats of bacteria to form domed deposits known as stromatolites. At Shark Bay the stromatolites cluster along the beach, generally within the intertidal zone. Intolerant of desiccation, they form squatter shapes in the shallows but may rise a metre or more from the sandy bottom where the water is deep enough. With an average growth

Stromatolites tend to be intertidal. Where they are close to the shoreline, many become completely exposed during low tide (ABOVE). On the other hand, their height is limited by the high tide, because the organisms which build them live only on the upper surfaces and cannot survive

rate of half a millimetre a year, the taller columns would seem to be several thousand years old, since their foundations are usually buried in the accumulated sediment of the bay floor.

Though Shark Bay holds the only major population of stromatolites left in the world, the cyanobacteria that build them live in many mineralised lakes, lagoons and springs. Only in the Bahamas however, do stromatolites approach the age and size of the Shark Bay population. Another large though little-known group of Australian stromatolites occurs along the eastern shores of Lake Clifton, 700 kilometres south of Shark Bay near the town of Mandurah, Western Australia. Unlike the massive domes and columns at Shark Bay, however, the stromatolites at all these sites are comparatively young and small.

All sites share four major characteristics: they are warm, sheltered, highly mineralised and support relatively few predators that graze on bacteria. Shark Bay predators are limited by the extreme salinity of the water, a result of minimal water circulation and a high evaporation rate.

Such warm, mineralised lagoons have always been common, especially on the margins of emerging land crust. By 2.8 billion years ago stromatolites had appeared in similar habitats throughout the world. A spectacular illustration of this first flourish appears on the banks of the

without regular immersion. Several species of cyanobacteria have been identified at Shark Bay and more than one species may coexist on the same stromatolite, building in their different styles. Larger, spherical species build irregular deposits that often look like cauliflowers (RIGHT).

STROMATOLITES, SHARK BAY, W.A.

Nullagine River, barely 80 kilometres south-east of the North Pole mining centre. The seething mass of bacterial life that once gathered there built an array of giant stromatolites which have no equal. Some are two metres high and almost as wide.

Fossil stromatolites, like modern ones, vary greatly. The differences appear to result from a complex interplay between the size, shape and feeding habits of their builders and their particular environment, which may range from the intertidal zone to moderately deep water. Several species of cyanobacteria may also share in the construction of a single stromatolite column, one species replacing another according to water depth, its turbulence and other factors. Each changeover is marked by a change in building style.

Stromatolites were a major component of early reefs, but unlike coral polyps, cyanobacteria have no skeleton to bequeath to their deposits. In fact, apart from the chlorophyll that now colours and powers them – and a single strand of DNA – cyanobacteria show almost no identifiable internal structure at all.

They share this lack of structure with all bacterial and viral organisms, and this primitive characteristic sets them apart from all

These smooth-topped stromatolites tend to crowd the shallows nearest the shores, and in some places their upper surfaces have merged to form a typically spongy platform that conceals a maze of small tunnels between their stems.

ZIPPER OF LIFE

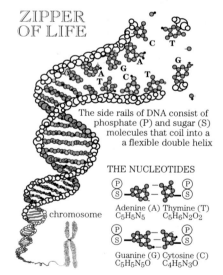

The side rails of DNA consist of phosphate (P) and sugar (S) molecules that coil into a a flexible double helix

THE NUCLEOTIDES

Adenine (A) $C_5H_5N_5$ Thymine (T) $C_5H_6N_2O_2$

Guanine (G) $C_5H_5N_5O$ Cytosine (C) $C_4H_5N_3O$

chromosome

All life seems to have stemmed from a single creative process. This would at least explain why all organisms are composed of the same few molecular components; why their main building blocks, amino acids, occur only in a special chemical form in living tissue; and why the heredity of every cellular organism on Earth is controlled by variations of the same thread-like molecule of deoxyribonucleic acid—DNA.

DNA provides the blueprint for all life. It carries the genetic information for the simplest organism to the most complex. While it is capable of infinite variety in the chemical sequences that make up its enormous length, its principal policy-making components consist of only four chemical bases, or nucleotides.

The whole structure is rather like a zipper that has been twisted into a spiral. Phosphate and sugar molecules reinforce each of the side rails, while four kinds of molecular teeth—the nucleotides—project towards each other from these sides. The teeth do not interlock as does a zipper, but each tooth bonds with the tooth opposite and only in the same chemical partnership every time, forming one of the two possible pairings.

In other words each chemical tooth can have only one kind of chemical partner on the opposite side of the zipper. Consequently, when DNA unzips to replicate itself during reproduction each side of the zipper forms a template of the other that is as accurate as the opposing edges of a torn page. Using this template, the cell may then rebuild new sides on the divided zip, producing two precise copies of the original DNA sequence.

This ability to replicate itself has been the key to the phenomenal success of DNA as the control mechanism for all life: it also provides strong argument in favour of such a molecule being one of Earth's first 'living' entities.

DNA may in turn have evolved from ribonucleic acid (RNA), a single-strand form of genetic material that now provides the communication medium within cells. It still runs some viruses.

STROMATOLITES, LAKE CLIFTON, W.A.

Australia's second largest population of stromatolites (*LEFT*) extends for almost ten kilometres along the eastern shores of a land-locked lagoon near Mandurah in south-western Australia. Little known and rarely studied, the builders of this group appear to be confined to a single species. In some places the mud is entirely sheathed in a thick mat of cyanobacteria and algae (*BELOW*).

STROMATOLITES, LAKE CLIFTON, W.A.

SIGNATURE

While the stromatolites of Shark Bay are by far the largest in the modern world they are small compared to some that grew along the shores of the Pilbara region some 2.8 billion years ago. Erosion by the Nullagine River has uncovered a layer of fossilised giants which, in many cases, measure nearly two metres in diameter (BELOW). Very different were the small branching forms (RIGHT) that grew in a central Australian seaway some two billion years later, while another kind entirely characterised ocean reefs about 360 million years ago (FAR RIGHT), just before stromatolites faded from prominence in the fossil record.

FOSSIL STROMATOLITES, PILBARA, W.A.

FOSSIL STROMATOLITES, MACDONNELL RANGES, N.T.

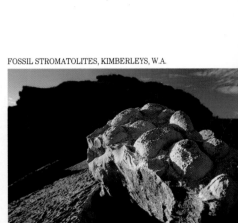

FOSSIL STROMATOLITES, KIMBERLEYS, W.A.

later forms of life. They are relics of a world whose atmosphere would have been alien and poisonous to us. But it was also these same tiny organisms which helped to change all that.

The explosive multiplication of such photosynthetic bacteria about 2.8 billion years ago released huge quantities of another waste product, oxygen. This was to endow Earth with its first energy resource that was biologically based, and evolution would find it irresistible. Powered by this stockpile of high-energy fuel, the main thrust of development would eventually swing from the staid economy of these bacterial primary producers to a group of oxygen-hungry parasites — animals. The whole animal world today runs on this same stockpile of bacterial oxygen, which is continuously recycled by modern photosynthetic organisms such as plants, algae and cyanobacteria.

In the barren world of 2.8 billion years ago, the profound diversity of modern life would have seemed improbable indeed. However, those simple blue-green cells were not only making such evolution possible by virtue of their oxygen waste, they were also burning the evolutionary bridges that lay behind them. The delicate, molecule-building processes which had created life's ingredients in the original oxygen-free atmosphere were about to become chemically impossible to repeat. Life could never begin all over again. Oxygen would 'burn' it up.

———

2.5 – 1.8 billion years

THE EARTH HAS CHANGED *little since life's first traces appeared some six hours ago. Almost half our time scale has already ticked away. Despite the enormous drag of its tidal seas, the planet is still spinning fast. Each day – one revolution – lasts only about 18 hours. The world is warm and cloudy and its surface is largely covered by shallow seas. Internal heat generated during Earth's birth processes continues to drive silicon to the surface. Like a crop of teenage boils, huge domes of silicon-rich magma form within the crust, crystallising as granite. Reinforced by this, land crust accumulates. 'Australia' at this stage consists of two major landmasses lying relatively close to each other somewhere near the north pole. There is at least one period of polar cooling during this time and ice appears briefly on Australia's Pilbara block. Its coastal waters are generally warm however, for they teem with bacterial life. Bacteria have proliferated to the point where their photosynthetic feeding processes have released sufficient of their waste gas, oxygen, to trigger a major environmental change. The iron that has been carried in solution in the seawater starts to oxidise. The seas begin to rust.*

These layers of Hamersley ironstone represent the greatest environmental crisis that life has ever triggered in its colonisation of the planet: its oxygen wastes 'rusted' the seas.

HAMERSLEY GORGE, HAMERSLEY RANGE, W.A.

BREATH OF LIFE

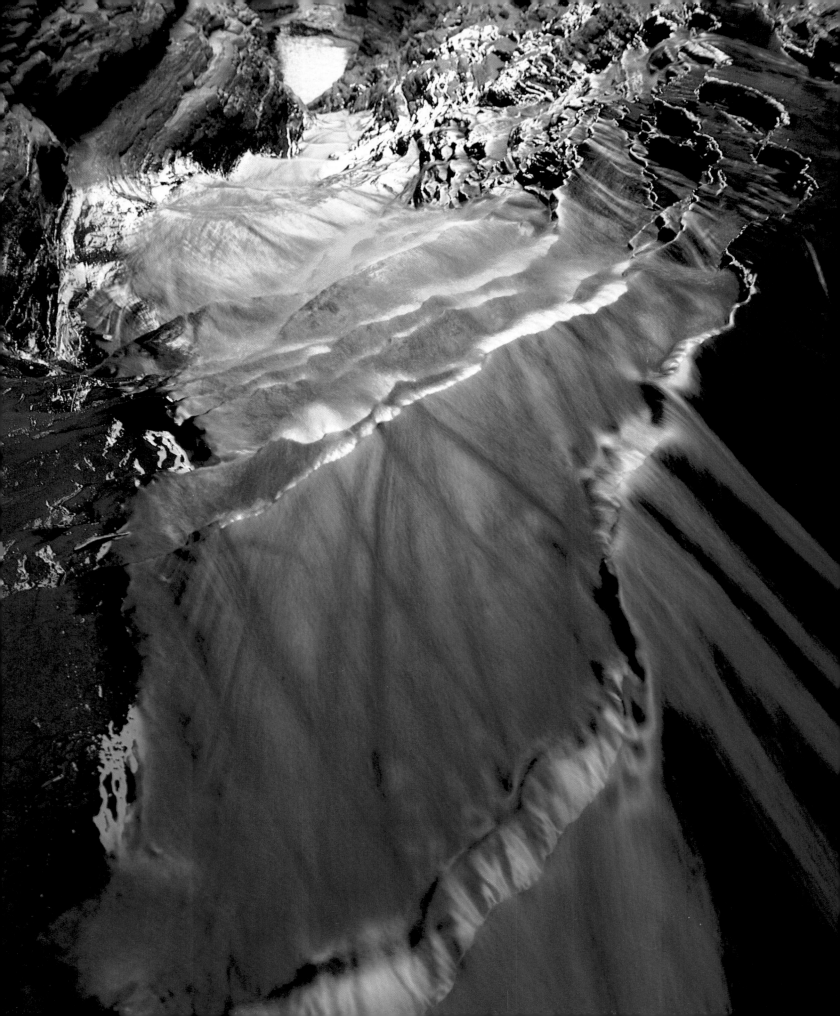

KNOX GORGE, HAMERSLEY RANGE, W.A.

BREATH OF LIFE

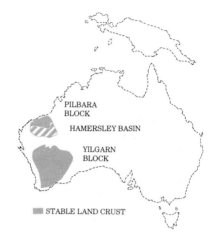

PILBARA BLOCK

HAMERSLEY BASIN

YILGARN BLOCK

STABLE LAND CRUST

The steady wear of wind and water on the iron hills of the Hamersley Range has spectacularly reshaped this remnant of life's greatest environmental catastrophe. These ancient Pilbara seabeds now incorporate the most colourful array of gorges in Australia, ranging from narrow water-worn corridors like Hancock Gorge (LEFT), to sheer-sided chasms such as Knox Gorge (ABOVE RIGHT). But while the gorges provide the spectacle, it is the iron oxide sediment from which they are carved that is most significant. This fallout from the rusting seas swept the water free of iron, allowing all subsequent oxygen waste released by the world's marine bacteria to leak into the atmosphere. This eventually turned our skies blue and provided the energy resource upon which most future evolution would depend.

WESTERN AUSTRALIA'S HAMERSLEY RANGE sprawls across the ancient Pilbara plateau like a tribal scar, raised and livid. Rivers have torn open its flanks and iron-rich gravel spills like dried blood over the plains. Dusted with green-gold spinifex and sprinkled with white-stemmed eucalypts, the Hamersleys display colour with a savagery that borders on the surreal. It is a landscape unlike any other on Earth.

A walk into one of the winding gorges does nothing to dispel the feeling of unreality. Smooth pavements and sheer walls, composed of neat rectangular ironstone blocks, create the illusion of a rusting city. Water slides across terraces and chatters down stone steps into deep, still pools, shaded by giant fig trees and native pines, and birdsong echoes eerily. Geologically the walk represents a step backward in time of almost 2.5 billion years. It is a journey that leads to the heart of a biological event that reworked the face of the Earth and changed the course of its history. The entire Hamersley Range is a memorial to life's first major confrontation with the finite resources of a small planet.

A billion years of molecular evolution – the weeding out by misadventure of unstable molecular unions – had produced a narrow range of cell-like organisms. Though their DNA was rudimentary and without the protection of any kind of nucleus, they survived as small, efficient chemical factories. They fed on a mineral soup of seawater and duplicated themselves when they grew too fat by a simple process of subdivision into more stable proportions.

The earliest forms were probably simple fermenters, harvesting natural carbohydrates such as glucose from the sea and breaking them down for their energy content. But as the Sun's growing radiation became more penetrating and Earth's storm clouds began to clear, some bacteria evolved a chemical process that used the energy in sunlight to internally synthesise the glucose nutrient from the more abun-

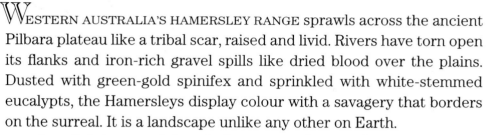

NP
60°
30°
0°
30°
60°
SP BILLIONS OF YEARS   *3*   *2*   *1*   PRESENT TIME

HANCOCK GORGE, HAMERSLEY RANGE, W.A.

A *walk into some of the deeper gorges in the Hamersleys represents a step back in time of over two billion years. One of the most spectacular is the Hancock Gorge which, in its lower reaches, winds crevasse-like between water-polished walls of high-grade iron ore.*

(OVERLEAF): This sediment represents hundreds of millions of years of rusty fallout from the iron-rich seas of the Hamersley Basin. It appears to have been triggered by a rapid rise in the rate of oxygen discharge by the microbial life that teemed along the Pilbara's southern shores about 2.5 billion years ago: more importantly, it probably signalled the first appearance of the chlorophyll molecule, or at least an early form of it. This complex and highly efficient molecule was to become the mediator of all photosynthesis, the process by which green plants trap the energy in sunlight and convert it to a usable form. Chlorophyll now underpins the existence of 99.9 per cent of all life on Earth.

HAMERSLEY RANGE, W.A.

The scale and pattern of recent erosion (LEFT) becomes more apparent from the air. This is the junction of Red Gorge and Knox Gorge.

The rusty cliffs (RIGHT), the white-stemmed Snappy Gums that cling to the them, and the blue, oxygen-charged skies that arch over them, all owe their existence to the same chemical process, photosynthesis — and probably to the same mediating molecule, chlorophyll. It was chlorophyll that powered most of the early oxygen-producing bacteria, and it remains the power source and colouring matter in all green plants.

dant simpler molecules outside. This process of photosynthesis was to become the blueprint for almost all future plant life.

The key to this explosive success lay in the construction of a complex protein pigment centred on a magnesium atom. The result was chlorophyll which, in its greenish form, colours most of the modern plant world.

We do not know when chlorophyll first appeared. The builders of the North Pole stromatolites may well have used other chemical pathways to feed. But it seems likely that the flourish of giant stromatolites that occurred about 2.8 billion years ago may well have signalled the perfection of the chlorophyll molecule. Certainly chlorophyll seems to have triggered the dumping of the Hamersley iron deposits some 300 million years later.

When chlorophyll appeared in bacteria their feeding became a two-step process that split the water molecules to extract the hydrogen needed for building glucose. The by-products were the limey minerals which had been dissolved in the water, and the other half of the water molecule, oxygen. This gas was highly poisonous in some forms but cells learned to cope with it, and the modification was an economic triumph. The energy production of each organism was increased some 18 times,

WEANO GORGE, HAMERSLEY RANGE, W.A.

HAMERSLEY GORGE, HAMERSLEY RANGE, W.A.

Late afternoon sunlight sends furnace colours echoing down the walls of Weano Gorge (LEFT). Despite meagre rainfall and long hot summers the deep rock pools scattered through these gorges never dry out. One of the most colourful of the Hamersley gorges is that carved by the South Fortescue River where it descends from the Hamersley plateau. Known as Hamersley Gorge, it is neither long nor deep by local standards, but aeons of water sculpture and a garish palette of ironstone pigments bestow a surreal quality on its common-place elements (ABOVE).

while its raw-material needs were simplified to carbon dioxide, water and light, all of which were plentiful.

Inevitably the new cells thrived and their durable rubbish dumps, stromatolites, suddenly became common in sediments younger than 2.8 billion years old. But in this success lay the seeds of disaster: though the new organisms could adapt themselves to the poisonous nature of their oxygen waste, they offered no defence against its environmental impact.

Massive erosion from the young landmasses had charged the sea-water with huge quantities of soluble iron. With the growing volume of free oxygen the seas began to rust, and a gentle rain of iron oxide sediment began to fall to the sea floor. In time this became a torrent that turned whole oceans red. Such was the case in the Pilbara where so many layers of iron silt were dumped by the coastal waters that they would one day form one of the richest ore-bodies in the world. Dried, compressed and eroded, the Hamersley group of sediments still measure almost 2.5 kilometres thick.

For the microscopic life that triggered this chain of events, the deposits of iron oxide came as an avalanche that threatened to bury it. By about 1.8 billion years ago, when the seas had eventually been swept clean of iron, much of the early life had disappeared with it. But in these less crowded waters more robust organisms appeared. They were oxygen producers also, and their gas wastes began to move freely from the sea into the atmosphere. This growing stockpile of aerial oxygen was destined to build the radiation shield of our blue skies and enable the crucial developments of sex and death upon which we, and the entire web of modern life, would depend.

The Hamersley Ranges remain a spectacular memorial to evolution's first major breakthrough. Yet though this breakthrough laid the environmental foundations for all future development, there was an ominous double edge to its message. The energy harvest reaped by those microscopic life forms exacted a massive fee, and buried much of the life that caused it. It was an environmental impact that would not be matched until the coming of Man.

2.3 –1.5 billion years

As the clock ticks *on into the afternoon of our time scale a subtle change in the atmosphere gradually alters the appearance of the sky. Part of the oxygen released by bacteria takes the form of ozone, most of which becomes concentrated some 30 kilometres above the Earth's surface. This ozone layer scatters the blue light it filters from the sun's spectrum and gives the sky a faint tinge of blue. It is the first visible sign of a radiation shield that will eventually allow life to emerge from the shelter of the sea. Australia's twin foundation stones, the Pilbara and Yilgarn blocks, have by this time returned to the mid latitudes and are heading south, accompanied by several new scraps of emerging land crust. Meanwhile, deep inside the planet a new pattern of convection is brewing. It heralds a massive shuffling of the Earth's crust and its surface scum, the continents. For the moment, however, the shallow seas that wash through the Australian region are warm and peaceful and alive with microscopic life. Drifting among clouds of feeding organisms is an occasional four-cell grouping that has a triangular arrangement inside a membrane sac. This is the hallmark of cells that possess nuclei. A genetic timebomb has been primed: all that is now needed to set it off is a suitable fuel. And the stockpile of oxygen is growing.*

The Devil's Marbles in central Australia are part of the continent's underlying skeleton. All its main component blocks are founded on granite platforms like this, and as their blanket of sediments grows threadbare with age, these bones begin to show.

DEVIL'S MARBLES, N.T.

BUILDING BLOCKS

McARTHUR RIVER, N.T.

BUILDING BLOCKS

WHEN MONSOON THUNDERHEADS GATHER along the Arnhem escarpment in the Kakadu National Park of the Northern Territory, its battered rock towers assume the gloomy grandeur of a medieval fortress under siege. It is an imagery that befits this relic of Australia's stormy past because its origins lie at the beginning of the most turbulent era in Australia's geological history, an era which laid the foundations of the modern continent.

Between 2.3 billion and 1.9 billion years ago, several huge slabs of new land crust began to emerge near the old continental cornerstone established by the Yilgarn and Pilbara blocks. It was also during this time that structural changes in the planet's interior altered the pattern of its convection currents: small, vigorous convection cells were replaced by much larger ones. On these floated a thin skin of semi-rigid oceanic crust, which was regularly recycled. Broken into a set of discrete plates, which moved independently like conveyor belts, this oceanic crust began to push its scum of continents all over the globe. This marked the birth of the modern system of tectonic plate movement. Australia's current rate of northward drift is about 5.5 centimetres a year. However, the average rate of crustal drift is no greater than the speed at which a fingernail grows, between three and four centimetres a year. Extended over three billion years, the age of Australia's oldest components, this amounts to a total journey of not less than 100,000 kilometres.

With the aid of modern dating techniques and sophisticated processes of geological analysis, we at last have the means to unravel a little of this story. The information is scrappy and offers little more than a series of 'postcards' left over from Australia's wanderings, but the scale is unmistakable. During the last three billion years Australia has visited

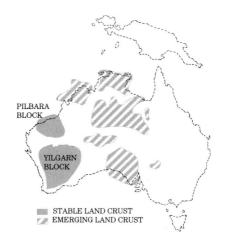

PILBARA
BLOCK

YILGARN
BLOCK

▨ STABLE LAND CRUST
▨ EMERGING LAND CRUST

The sand ripples (LEFT) were probably formed in the shallows of a large freshwater lake, more than 1.5 billion years ago. They now lie in Jim Jim Gorge at the foot of the Arnhem escarpment.

Spectacular gorges (ABOVE RIGHT) are not confined to the western escarpment of the Arnhem plateau; the McArthur River has carved this sheer-sided gorge through the pink sandstones that fringe the south-eastern edge of the Arnhem block.

(OVERLEAF): The rain-swollen rivers that plunge off the edge of the Arnhem escarpment during the monsoon have carved out innumerable deep gorges, filled with water.
Among the most scenic of these gorges is the one gouged into the escarpment by the Jim Jim River at the southern end of Kakadu National Park.

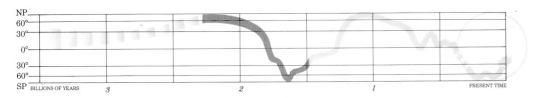

NP
60°
30°
0°
30°
60°
SP BILLIONS OF YEARS   3   2   1   PRESENT TIME

BUILDING BLOCKS

As blocks of young land crust (BELOW) entered the final stage of their development they often became flooded with sheets of lava. This boulder field near Mount Herbert in the Chichester Range is the remains of such a lava flood that covered the Pilbara block with basalt about 2.7 billion years ago, just before the region stabilised.

Scraps of the original granite show through (RIGHT) where the sandstones of the Arnhem plateau have eroded away. This outcrop, beside the South Alligator River, is known to its Aboriginal owners as Monitor Lizard Dreaming.

CHICHESTER RANGE, PILBARA, W.A.

MONITOR LIZARD DREAMING, KAKADU, N.T.

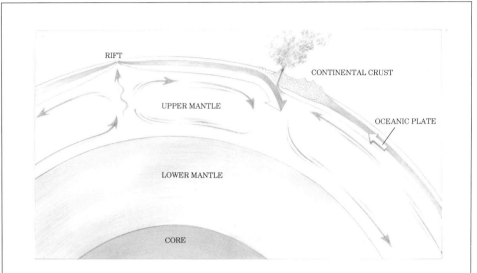

SCUM ON THE MELTING-POT

The Earth's surface consists of two kinds of material: oceanic crust, formed like a skin on the semi-molten rock of the upper mantle; and continental crust, a buoyant and permanent 'scum' embedded in this skin. Driven by huge convection currents that flow beneath it, the oceanic crust is broken into a number of separate plates that constantly recycle. Where plates are pulled apart new material is added, while at the opposite edge, old material sinks back into the melting pot. As they ride on these crustal conveyor belts, continents become redistributed whenever there is a change in the flow pattern of the mantle.

both poles twice and grown from two small crustal blocks into one of the world's major continental units.

The north-south movements of continents may be deduced by comparing the variation between patterns of contemporary magnetisation that become imprinted in some rocks during their formation. The oldest evidence of this fossilised magnetism yet found in Australia occurs in the Pilbara in lava that crystallised 3.45 billion years ago. This suggests that Australia then lay somewhere north of the equator.

BUILDING BLOCKS

'STONE FOREST', N.T.

Erosion eating into vertical joint lines at the southern edge of the Arnhem plateau has dissected this sheet of shallow marine sediment into a forest of stone pillars. Most of them are tightly packed and up to 20 metres high, making this one of the most remarkable landforms in Australia.

DISSECTED SANDSTONE, McARTHUR, R., N.T.

By the time the present system of tectonic motion became established 1.5 billion years later, the Pilbara and Yilgarn blocks had been to the north pole and back, a journey of 10,000 kilometres or more. Barely 250 million years later the Australian complex, which included the emerging landmasses that now form the Kimberleys of Western Australia, the northern half of Northern Territory and most of South Australia, lay beside the south pole, almost 15,000 kilometres away. Continental drift seems to have accelerated enormously.

Caught in the Herculean tides of these moving plates, the young land crust that formed these blocks was time and again deformed and fractured. Massive injections of molten granite about 1.8 billion years ago marked the onset of rigidity between 1.7 billion and 1.4 billion years ago, allowing erosion to draw a shroud of debris over their scarred remains. The battered remnants of these overlying sediments now provide much of northern Australia's most rugged scenery.

The Kimberley and Arnhem blocks are well exposed. Others have

The chocolate-coloured basalt that crowns the graceful buttes and mesas of the Chichester Range (LEFT) was once part of a vast sheet of volcanic lava that flooded over the southern Pilbara as the landmass stabilised. One of the best known features in the region is the elegantly shaped butte known as Pyramid Hill (RIGHT).

PYRAMID HILL, CHICHESTER RANGE, W.A.

been so eroded and buried that little is now visible. The Davenport Range south-east of Tennant Creek, the granite tors of the Devil's Marbles, and The Granites, are all relics of this period. Likewise, little is visible of the massive Gawler block that underpins most of the Nullarbor Plain and the Eyre Peninsula. The tangle of mineral-rich hills around Broken Hill and Mount Isa are also iceberg-like tips of outlying crustal blocks that originated during this period.

These continental building blocks, which comprise two thirds of the modern Australian landmass, are monumental indeed. Buried in the sediments that cover these blocks are other mementos of great significance to us but their historical dimensions become clear only under a microscope.

Among the fractured ranges that run northward from Halls Creek is a small pile of old seabeds. They have been gracefully scalloped by erosion to reveal a layer-cake of sediments interleaved with dolomite, a kind of limestone. The dark layers of dolomite are the biological wastes precipitated by microscopic life that teemed in those shallow seas between 1.6 billion and 1.7 billion years ago. Among the organisms that became fossilised in the limey ooze of these shallow seas were occasional clusters of four microscopic spheres locked together in a triangular pyramidal formation. In this arrangement, known as a tetrad, each sphere touches the other three. Similar microfossils occur in limestones and shales of the same age some 700 kilometres to the east, near the Gulf of Carpentaria. A few tetrad cell clusters found there were also enveloped by an outer membrane.

Many modern cells display both these features – the pyramid cluster and outer membrane – during certain stages of development. Such cells always share two other characteristics: they have just undergone a reproductive division, and much more significantly, each has its DNA coiled within a small membrane sac, a kind of cell within the cell. In other words, each has a nucleus.

These fossils represent the greatest single evolutionary leap ever taken by life on Earth. Here was a developmental watershed that

The seabeds that form these layercake hills (ABOVE) in the eastern Kimberleys teemed with bacterial life abaout 1.6 billion years ago. Among the host of microfossils that they contain are a few frozen in an act of reproduction that fits no bacterial pattern. It seems instead to match the system used by our own body cells during non-sexual reproduction, a system known as mitosis. The difference is that mitotic reproduction is only available to cells which have a nucleus. The development of the nucleus marks the greatest watershed in all evolution.

(ABOVE RIGHT): The sandstone rock stacks of the Tabletop Range south of Darwin were laid down at the same time as the Arnhem sediments, about 1.5 billion years ago, and they represent the western edge of the same landmass. The region has been relatively stable ever since.

BUNGLE BUNGLE DOLOMITE, KIMBERLEYS, W.A.

TABLETOP RANGE, N.T.

irrevocably split evolution into two separate streams: one consisting of unnucleated organisms known as prokaryotes, and one in which the cell's genetic material was packaged within a membrane sac. These were the eukaryotes.

It is uncertain exactly where or when this evolutionary watershed occurred but it probably owes its origins to the massive pollution of the world's oceans by the bacterial waste gas, oxygen, some 2 billion years ago. The internal structures, or organelles, that characterise all modern eukaryotes are thought to be remnants of predatory bacterial invaders. Foiled perhaps, by their host's well-packaged DNA, they seem to have remained within the cell and evolved a symbiotic relationship with the host's genetic material. Inherent in this biological complexity was the seed of another crucial evolutionary innovation – natural death. The human intestinal bacterium *Escherichia coli* divides every half hour. Theoretically, this single prokaryotic organism could, within three days, reproduce a mass of cloned replicas that would equal the weight of the planet. Success on such a scale had to be curbed. Each nucleated cell now incorporates an efficient self-destruct mechanism that is based on a process of genetic error accumulation. After a certain number of divisions the eukaryote develops irregularities and dies.

This process has never been nature's normal method of terminating an animal's life, but it serves to increase its vulnerability to predation, disease and various physical misadventures. Cocooned by technology, modern humans may avoid 'the thousand natural shocks that flesh is heir to', but the error clock ticks on within our cells, making each a tiny time bomb and death inevitable.

1.6 – 0.9 billion years

Wᴵᴛʜ ᴀʟᴍᴏsᴛ ᴛᴡᴏ ᴛʜɪʀᴅs *of our time scale elapsed, the Earth enters the mid-afternoon of its existence. Climates are generally warmer than today. There is no sign of ice, even at the poles, and the seas are teeming with microscopic life forms. Their oxygen waste by this time has increased, but it would constitute little more than two per cent of today's atmospheric level. Three large Australian landmasses, having passed across the south pole, are drifting slowly northward into the tropics once more. Each slab of land crust is fringed with an apron of thinner, more pliable material. This provides a buffer zone between them, which crumples into mountains or sags into the sea, according to the jostle of the crustal plates. In the compressed time frame of our 24-hour scale, these localised disturbances ripple through the buffer zones with the brief intensity of summer storms, leaving mountainous folds, fractures, sheets of lava and old seabeds strewn in their wake. Continental drift carries the Australian landmasses to a collision at the north pole. It results in the birth of the modern continent and the start of a billion year association with Antarctica. This bond will eventually cast Australia in its modern role as the last refuge for Antarctic life – a great southern ark.*

This jagged wall of silica marches in dislocated segments over the hot, stony hills that lie at the south-eastern corner of Western Australia's Kimberley region. As a silica filled fracture line, it traces one of the innumerable faults that characterise this belt of much-deformed crust, which bonds the Kimberleys to the rest of Australia.

CHINA WALL, HALLS CREEK, WA.

WELDING

STANDLEY CHASM

WELDING

MOST OF AUSTRALIA IS NOW VEILED by a sheet of windblown sand. It drapes in broad swathes about the old crustal blocks on the western half of the continent, concealing the weld lines that bond them into a unit. For a billion years these joints served as buffer zones. Their thin, mobile crust absorbed the titanic stresses of continental birth by contracting into mountainous folds during periods of compression and at other times sagging to form long, shallow seaways. The wrinkles and scars left by these events may be sparse but even in their understatement a monumental scale is discernible: such are the character lines in the face of this ageing continent.

The first tremors of unrest ran through Australia's internal buffer zones about 1.9 billion years ago. The welding began in earnest 300 million years later, though it took a further 700 million years before the three continental slabs, or cratons, bonded into a unit. The western two-thirds of the continent approximated the form we know today, but there was one major difference: Australia was not alone. Along its 'southern' borders lay the shadowy mountains of Antarctica. This was the beginning of an association that would last for more than a billion years and help to shape both the form and distribution of much of Earth's future land life. It was this association with Antarctica that would eventually cast Australia in its modern role of a southern ark, ferrying a cargo of unique Gondwanan lifeforms into the future.

The origins of Australia's collision with Antarctica have become blurred during vast periods of time. There are signs of compression along the southern edge of Western Australia that appear to be at least 1.6 billion years old, though the cause is as yet untraceable. A major compression and continental welding did occur, however, between 1.3 billion and 1.1 billion years ago, and there is evidence of this in Antarctica. The sequence of deformations imprinted in rocks in the Bremer

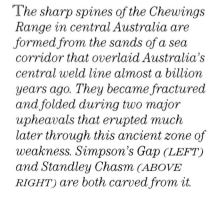

The sharp spines of the Chewings Range in central Australia are formed from the sands of a sea corridor that overlaid Australia's central weld line almost a billion years ago. They became fractured and folded during two major upheavals that erupted much later through this ancient zone of weakness. Simpson's Gap (LEFT) and Standley Chasm (ABOVE RIGHT) are both carved from it.

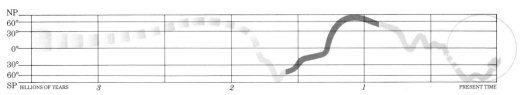

Bay region of Western Australia are unmistakably mirrored 2,000 kilometres away, in the Windmill Islands, near the Australian Antarctic base, Casey.

The cold swells of the Southern Ocean now conceal most of the weld line that bonded these two coastlines for so long but the grey granite domes, once molten, which now rise out of the sea along Western Australia's southern edge, are reminders of the enormous forces that were expressed here during the continental welding.

By contrast, Australia's internal sutures have held firm. These belts of thin crust that separated the older continental blocks frequently lay

The gently rounded granites of William Bay (BELOW) give little hint of their turbulent origins. They were born far underground in a cauldron of molten rock, generated during the titanic jostling that resulted in the billion-year connection between Australia and Antarctica.

WILLIAM BAY, W.A.

EAST MOUNT BARREN, W.A.

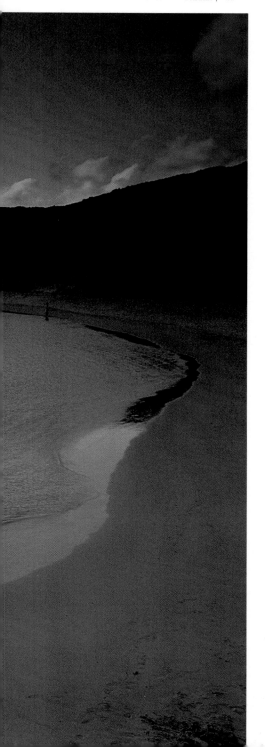

NATURAL BRIDGE, ALBANY, W.A.

These former sandstones (LEFT) were fused and deformed by the massive heat and pressure generated during the collision that bonded Antarctica to Australia about 1.1 billion years ago.

This bridge of granite (LEFT) near Albany, in south-western Australia, was once part of the 'glue' that bonded this section of coastline to a portion of Antarctica that lies to the west of the Australian base of Casey. The sea began to carve features such as this when it swept back into the weld line as the two continents began to separate, some 50 million years ago.

Still looking a little like the boils of molten rock they once were, the islands of the Recherche Archipelago (BELOW) are the tops of a chain of hills that sank into the sea when the two continents finally pulled apart.

RECHERCHE ARCHIPELAGO, W.A.

CHAMBERS PILLAR, SIMPSON DESERT, N.T.

MACDONNELL RANGE, N.T.

GARDINER RANGE, N.T.

NORTH WALL, KING'S CANYON, N.T.

CHARACTER LINES OF A CONTINENT

It took more than a billion years for the western half of Australia to weld into a unified slab. The belts of material that filled the joints and formed the seams remained zones of weakness for a long time afterwards. These seams, thin and mobile during much of their development, frequently sagged below sea level. This was particularly true of the broad belt of mobile crust that ran through central Australia. As a result this region is now characterised by many layers of shallow-water marine sediments, the folds and fractures having been caused by crustal readjustments that continued to find the least line of resistance there long after the rest of the continent had stabilised. This whole wrinkled region now reflects the turbulence of Australia's formative years, like character lines in a human face.

SEX – THE BOUNTIFUL BLUNDER

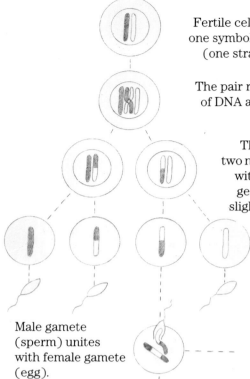

Fertile cell, showing nucleus and one symbolic pair of DNA strands (one strand from each parent).

The pair replicates, and sections of DNA are exchanged during a 'cross-over' phase.

The cell divides, forming two non-identical cells, each with a full complement of genetic material but with slightly altered sequences.

Both cells divide, forming four sex cells, or gametes, each containing a single, unique copy of the DNA code (in this case male).

Male gamete (sperm) unites with female gamete (egg).

Female parent

Male and female DNA codes pair, a nucleus forms, and the new generation begins.

It seems likely that sex is based on an old genetic blunder, a recurring copy error in a batch of flawed DNA, which eventually became institutionalised and highly specialised.

Even today the process begins like a normal cell division—mitosis—which typically produces two cloned cells. Instead, the final product is four sex cells, or gametes, each containing half the normal genetic information and each one entirely unique. In this process of reproduction, known as meiosis, the parent DNA passes through eight major phases, including a minor shuffling of some gene sequences and two different kinds of cell division.

Such a complex and vulnerable process would seem to have been doomed to failure, but for two peculiar advantages. Less viable sex cells were quickly weeded out, since any genetic weakness in their half-code was readily exposed by the difficulties of finding other compatible gametes in the hazardous environment. Secondly, the fertilised cell enjoyed the decisive advantage of inheriting slightly different genes from each of its parents. The presence of pre-tested alternative genes not only built in a degree of individual variability, it also short-circuited the genetic error-accumulation carried over from their parents' ageing processes.

In other words sexual reproduction offered the individual gene its best possible chance of survival during its journey into the future.

With optional genes built into the process, the chances of producing an organism suited to a particular environment were considerably enhanced. Meanwhile genes that were by-passed and unexpressed in the construction of one body could still travel on through time, dormant through countless generations if necessary, to a reawakening that might lie thousands, even millions of years in the future. In this sense the world is inhabited not by organisms but by their genes. The bodies in which they live are merely disposable vehicles, built by gene communities to ensure a safe passage through the hazards of a particular environment.

For the single cell, however, sex was a costly asset in its original form. Relying on chance matings in open water, such a system required a huge outlay of expensive genetic material to ensure a single union, let alone a reproduction rate that promised species' survival.

This wastage demanded a massive investment of energy and materials by parent cells. But with oxygen levels still rising, sex was here to stay.

Bacteria are able to achieve an extraordinary range of variation by a promiscuous exchange of genetic material. Most simply shed copies of their genes into the surrounding medium and, in turn, incorporate genes shed by other bacteria—even unrelated forms.

However, oxygen threatens the integrity of such unprotected genetic material, and as oxygen levels rose in primitive seas, this haphazard process would have become much less viable.

Sexual reproduction, on the other hand, provides genes with an efficient packaging service and an armoured delivery vehicle—the gamete or germ cell. The system may be both inefficient and uneconomical at a molecular level, but it offers a degree of genetic stability and reliability that bacterial gene-exchange cannot match.

BITTER SPRINGS LIMESTONE, MACDONNELL RA., N.T.

The slabs of sculptured limestone (RIGHT) that rise among the quartzite ridgelines of the eastern MacDonnell Ranges contain the remains of the teeming sea life that thrived in Australia's central seaway about 900 million years ago. Among the large, complex cells that characterise it are some of the earliest signs that sexual reproduction had evolved.

Evidence of this remains in the form of a Y-shaped stigma embossed on the thick walls of several cells fossilised in the limestone seabed that now lies exposed at Bitter Springs. This stigma is the scar formed at the junction of a tetrahedral group of four sex cells, or gametes, during meiotic division.

Only gametes display such scars.

below sea level, and the history of the central mobile belt is essentially that of a shallow sea corridor. The most graphic display of this lies along its northern margins where, fused by heat and pressure, the soft sands that once lined this central seaway now rear out of the modern sand-plains in towering walls of iron-stained quartzite. Simpson's Gap and Standley Chasm, two of central Australia's more spectacular gorges, are carved from this quartzite.

Further to the north other old seabeds emerge as a row of peaks that are among central Australia's highest – Mount Liebig, Mount Edward, Mount Zeil, Mount Sonder and Mount Giles. At the southern edge of the central mobile belt the Tomkinson Range, Mann Range and Musgrave Range also include seabeds of this age. To the east and west, however, the mobile belt submerges beneath the sands of the Simpson Desert and the Great Victoria Desert. This conceals all but the merest traces of the two big western arms which enfold the Pilbara-Yilgarn craton.

After a billion years of jostling, tectonic peace finally returned to these internal mobile belts around 900 million years ago, when the three major cratons bonded into a unit. But the seas that then washed in through central Australia overlaid that geological peace with biological revolution. It arrived in the form of a new kind of cell which bore a Y-shaped stigma embossed on its fragile casing. This, a birthmark of its peculiar reproductive process, remains a stigma of great significance for us all, for we too bear the same scars. The only kind of modern cell that shows this marking is one which has just been produced by a specialised system of cell division known as meiosis. It is the basis of the bizarre reproductive process we call sex.

950 – 600 million years

W**ITH MORE THAN THREE QUARTERS** *of our 24-hour time scale elapsed, it is well into the evening of our day. Earth's major landmasses have congregated in the northern hemisphere and Australia, bonded to Antarctica, is among them. The aerial stockpile of free oxygen is approaching three per cent of its present atmospheric level. In the sea, some cells have already begun to use oxygen to power new processes of feeding and reproduction. These developments fuel an explosive acceleration in the evolutionary process. Large, complex cells begin to roam the seas.*

Unified and stable at last, Australia turns southward once more, just as climates everywhere begin to deteriorate. It is the beginning of the world's most massive and least accountable climatic disturbance. It occupies two hours of our time scale and no major continent escapes unscathed. Glaciation is widespread but episodic, and it is severest in the tropics rather than near the poles. The first ice sheets form in Africa at about 7 p.m. on our scale. Glaciers do not gather in Australia for another hour but huge ice sheets then ravage the continent twice.

When the world returns to normal, Australia will have been ground flat and the life that crowded its coastal waters will have been decimated. It is a major episode of extinctions but it clears the way for a burst of genetic extravagance that is awesome.

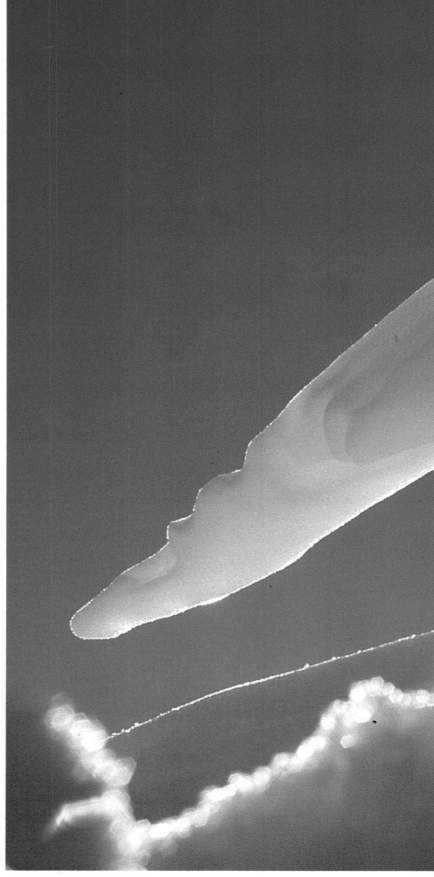

SNOWFORM, SNOWY MOUNTAINS, N.S.W.

AGES OF ICE

TILLITE, INDULKINA RANGE, S.A.

AGES OF ICE

950 MILLION YEARS TO 600 MILLION YEARS (1902 HOURS – 2052 HOURS)

PROBABLE
ICE SHEET   SEA

EDIACARA

ANTARCTICA

*The two boulders (LEFT AND
ABOVE RIGHT) that lie frag-
menting in the desert sun of
central Australia arrived there
stuck to the bottoms of glaciers
some 750 million years ago.
The soft siltstone in which they
now lie was then the mud bed
of a central Australian seaway
that bisected a continent
entombed in ice.
Outcrops of glacial debris are not
common in central Australia but
below ground the belt extends
almost unbroken from the
Flinders Ranges, in South
Australia, to the Kimberleys.
Even more remarkable is the
evidence that Australia then lay,
not in polar regions, but in
the tropics.*

SUMMER LIES HEAVY on the arid interior of the modern Australian continent: in the heat of the day miraged hills rise like pink scones from the
grilling plains. Birds and lizards sit agape. Here winter's chill seems
beyond imagination.

As you toil up the ridgeline towards the dark cliffs of Chambers
Bluff in the Indulkana Range, near the northern border of South Australia, an occasional rounded boulder jars the senses. Their smooth outlines and colour variation are in marked contrast to their geological sur-
roundings, a region of hard-edged, heat-shattered rock. The rounded
rocks of granite, quartzite and limestone signal the outcrop of a gigantic
belt of ancient glacial debris that lies buried far below the arid horizon.
This bed stretches almost unbroken from the Kimberleys in the far
north-west of Western Australia, to the Flinders Ranges in South Australia. The mud, sand and boulders that filled this glacial rubbish dump
were ripped from far-off southern mountains by the first sheets of ice
that closed over Australia some 750 million years ago, at the height of
Earth's most devastating Ice Age.

It was the first of two major onslaughts on Australia that were
unleashed 80 million years apart. They were part of a series of climatic
aberrations that sprawl across more than 400 million years of the
planet's history. Every major continent bears the scars of glaciations
that seem to have occurred during this period.

What triggered such massive cycles of cold is not certain. There is
fragmentary evidence scattered around the world of a much earlier ice
age, around 2.3 billion years ago, but it remains a shadowy event of
unknown dimensions. There have been several others – we are in the
grip of one now – but none compare, either in duration or ferocity. Australia was not even in the polar regions at this time of terrible cold.
According to the evidence of magnetisation patterns that have been

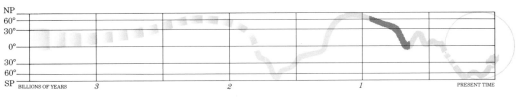

NP
60°
30°
0°
30°
60°
SP
BILLIONS OF YEARS   3   2   1   PRESENT TIME

TILLITE, FLINDERS RANGES, S.A.

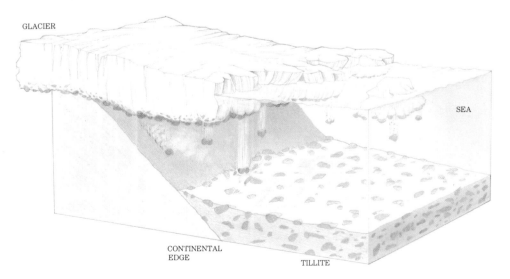

GLACIER

SEA

CONTINENTAL
EDGE

TILLITE

In Tillite Gorge, in the Flinders
Ranges (ABOVE), boulders of all
kinds are scattered through the
silty matrix, like currants in a
bun. At the northern end of the
Flinders the same deposit of
glacial debris reaches a thickness
that is more than half the height
of Mount Everest.
Glacial rubbish dumps like this
usually form where an ice sheet
first meets the sea. (LEFT).

POLAR ICE: CLOAK OF MYSTERY

Earth's polar regions have been free of ice for most of the planet's existence. The current icecaps are certainly not exceptional, however, similar ice sheets having appeared many times before.

Many periods of global cooling are recognisable in the geological record and there have been at least five major sequences of glaciation. The causes of such episodes of cold are not yet well understood. There are signs of several component cycles that affect the Earth's temperature budget, yet the growth of polar icecaps seems to hinge more on random factors.

Most of the heat that drives the world's weather patterns comes from the sun. This incoming energy is not constant and it is modulated by an array of underlying pulses which vary in length from 11 years (the 'sunspot cycle') to one which spans some 280 million years (the period our galaxy takes to complete a revolution). Cooling increases when any of these cycles coincide.

Other variables that modify incoming energy are the Earth's distance from the sun, its speed of daily rotation and the tilt of its axis. In the planet's own area of influence, the final heat balance is limited by the composition of the atmosphere, which both filters the incoming rays and then traps reflected heat in what is known as the greenhouse effect .

Ultimate heat control seems to lie deep within the planet itself, where the convection currents of the mantle shuffle the continents about the crust. Land reflects much of the sun's energy, warming the air above it. The

SUN ON SNOW, FALLS CREEK, VIC

oceans, by contrast, form a 'heat sink', and readily absorb almost all the incoming infra-red rays. Thus when continents cluster in polar regions they open up large tropical oceans, minimise the heat reflection from the land and provide a stable platform for the accumulation of snow and ice. An ice age seems to occur when such random cooling factors combine with one or more cycles of reduced solar input.

About a billion years ago, however, the Earth entered a sequence of ice ages that broke all the rules. Their signatures are scrawled across 400 million years of geological record and are registered on most of the world's major landmasses. Even more intriguing is the evidence that all the glaciations seem to have occurred in tropical or sub-tropical regions, nowhere near the poles. Australia was engulfed by ice at least twice,

once as it drifted south across the Tropic of Cancer, about 750 million years ago, and again as it lay astride the Equator, some 80 million years later.

No theory adequately accounts for all aspects of these events but the simplest conclusion is that glaciation was in fact global and the apparent tropical bias is merely a product of inadequate evidence. One spectacular alternative suggests that the Earth became tipped on its side at this time during a collision with another massive object. Like a spinning top that has been knocked off balance, the Earth would have taken a long time to recover from such a blow. Its effect on world climates would have been catastrophic.

Earth's present orbital plane is slightly inclined from that of the sun's equator, and the 23.5 degree tilt on its rotational axis also points to such a collision having occurred at some time. But the question remains unresolved.

[Ice Age: Unfortunately three different meanings are attached to the term, 'Ice Age'. It is frequently applied to a single glaciation, lasting only a few thousand years, and sometimes it is used to refer to an entire glacial era—a succession of glaciations bounded by two major intervals of relative warmth known as interglacials. Occasionally it is also applied to the complete sequence of glacials and interglacials, which characterise an entire period of global cooling. In this text it is used only in this largest sense, since there is no convenient alternative term for it.]

TILLITE, KIMBERLEYS, W.A.

DROPSTONE, INDULKANA RA., S.A.

DROPSTONE, EAST MACDONNELL RA., S.A.

Massive beds of glacial debris (ABOVE LEFT) sprawl across many parts of the southern and eastern Kimberleys. These cliffs, exposed by erosion in the Ord River Basin are part of a 750 million year old deposit that appears to have been dumped by glaciers from the north.

There are also several places in the arid heart of Australia where you may stand on a sea floor of this age amid ice-polished boulders of many kinds. They were dumped there from the undersides of glaciers and icebergs as they melted in the seaway that bisected the continent at that time.

'fossilised' in Australia's rocks, the continent lay somewhere in the northern tropics.

Evidence of the ice is unmistakable. During its journey from snow-fields to the sea the sole of a glacier becomes impregnated with a mat of boulder rubble, turning it into a gigantic sanding belt. Even the hardest rock lying in its path is scoured and polished, while softer rocks are ground to powder. When a glacier reaches the sea, huge slabs break off and float away as icebergs. As these melt the rocks fall, littering the muddy sea floor with what are known as dropstones. It is a portion of just such a seabed that surfaces at Chambers Bluff, in the Indulkana Range. Though outcrops are rare, these layers of glacial debris, called tillite, lie beneath much of central Australia: they fall within a belt that runs from the Kimberleys to the Flinders Ranges – the old mobile belt. The weight of the ice sheet that rested along the southern half of Australia was so great that it depressed the entire southern side of the continental raft. This was sufficient to reopen the old central seaway, into which the glaciers dumped their rubble. At the northern end of the Flinders this glacial rubbish dump is a stunning 5.5 kilometres thick. One of the best outcrops occurs in Tillite Gorge near Arkaroola.

It is the sequel to this glaciation that is of peculiar importance to us, however, an event that began as the glaciers withdrew. As temperatures and oxygen levels rose, a wholly new breed of cells appeared. They gathered in formalised cooperative groups, each of which lived, or died, as one. These were the first animals in the common sense of the word, that is a corporate body of cells which feeds upon other organic material without photosynthesis. Proof of the existence of these animals was first found near the Flinders Ranges, in the seabed sequence that begins just above the tillite.

During the next 600 million years these small beginnings would give rise to some 1.5 billion species.

650 – 450 million years

THE EARTH HAS JUST ENTERED *the 21st hour of our time scale. Water has been available for about 18 of those hours and life has existed for at least 17 of them. But no multicelled animal has yet appeared. Clearly the miracle of life lies not in the creation of cells but in their cooperation.*

World climates are recovering from the worst ice age of all time, and earlier life forms have been decimated. The ice has barely withdrawn, however, when new kinds of cells living in highly specialised cooperatives scatter their signatures along Australian shores. These delicate impressions, left by some of Earth's first multicelled animals, will be preserved with eerie fidelity for almost 600 million years. Complex and elegant, they burst into the geological record with an abruptness that is breathtaking, like flowers born without buds. Some of these experimental forms will fail, never to reappear in the fossil record: others will provide the bloodstock for most of the world's future animal life, ourselves included.

The delicate impressions of large multicelled animals, like this segmented sheetworm, burst into the fossil record in the same period around the world. They were first discovered at Ediacara, near the Flinders Ranges, South Australia, where this elegant specimen was found.

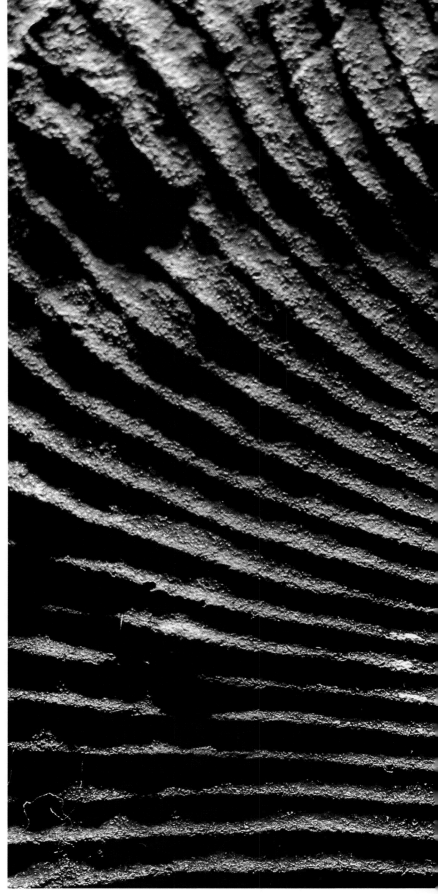

Dickinsonia costata, EDIACARA, S.A.

CORPORATE BODY

FOSSILS, EDIACARA, S.A.

CORPORATE BODY

LOOKING NORTH FROM the windy rim of Wilpena Pound, the neatly scalloped ridge lines of the northern Flinders Ranges curve away into the heat haze in almost military order. From here a sweep of the eye may traverse in a moment the massive products of almost a billion years of crustal events. Yet hidden among these crumpled seabeds are the traces of another event, one which far outweighs them all. By comparison its symbols are as delicate as snowflakes: each marks the death of a diaphanous marine organism.

These fossils are among the very earliest traces of multicelled life anywhere on Earth. A South Australian geologist discovered them in 1946 while working in the Ediacara Hills beside the Flinders Ranges. Yet his discovery earned him only disbelief. The variety, size and complexity of the impressions, allied to their apparent lack of antecedent forms, all helped to evoke a degree of scepticism that went beyond the call of scientific duty. 'Ediacaran' fossils have since been recognised in many parts of the world. Yet no trace of ancestral forms, either animal or protoctist (a probable precursor), have been found.

There now seems to be a simple answer. Ancestral forms existed, but because they lacked a single factor, the fibrous protein collagen, in death they would have left no trace. Animals are unable to construct collagen at oxygen pressures lower than about three per cent of present atmospheric levels (PAL). Without this stiffening material, their soft bodies would have disintegrated as soon as they died, leaving no imprint in the sediments that buried them. Atmospheric oxygen seems to have crossed the 'collagen boundary' about this time.

Some Ediacaran forms were quite complex, and among these was a broad, flat segmented organism called *Dickinsonia*. The key to its immediate success lay in the degree of specialisation and cooperation that its member cells achieved. In return each cell gained unparal-

PROBABLE
ICE SHEET

SEA

EDIACARA

ANTARCTICA

Some early experiments in body design were so effective that there has been little change to them in 600 million years. The common bluebottle (LEFT), so often washed up on Australian beaches, is typical. It is, nevertheless, just a cooperative of specialised cells, not a corporate body of them.

(ABOVE RIGHT): Fossils such as those embedded in Ediacara's hills were considered improbably large and complex for the age of the rocks that bore them. Even more perplexing was their lack of antecedent forms. It is now certain that they did have predecessors, but because these creatures lacked the gelatinous protein collagen, their shape was not retained after death, denying history a fossilised record.

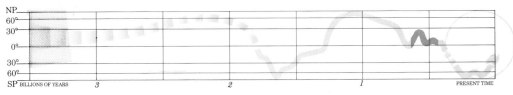

NP
60°
30°
0°
30°
60°
SP BILLIONS OF YEARS   3   2   1   PRESENT TIME

STINGING CELLS, *Physalia physalis*, QLD.

leled security and greater efficiency within the constant environment of the body corporate. But there was a price. A large body needed more fuel – both food and oxygen – to function, and failure, like success, was now a community affair.

Circulatory systems were in their infancy and cells still relied on direct absorption to obtain their oxygen, so the first animals had to be thin, deeply segmented, or very small. Such shapes exposed the greatest possible surface area to gas exchange. Some *Dickinsonia* grew almost a metre long, nearly as wide, yet no more than three millimetres thick when fully expanded. They probably doubled their thickness during contraction. Even this very limited mobility was costly, however, because it required the use of muscle cells, and consequently more oxygen.

To power these contractile cells, animals had added a new 'afterburner' stage to their normal fuel cycle. This unleashed 18 times more energy than the simple splitting of the glucose molecule. The source of this new power lay within the oxygen-loaded tail of a molecule known as adenosine triphosphate, or ATP. Easily transported and stored in muscle cells, ATP was readily broken down to achieve an explosive release of energy. Sleek, power-laden muscles still drive the complex machinery of the whole animal world with this same molecule. We call the process respiration.

The first flourish of multicelled life ushered in a period of unfettered genetic experiment in which an astonishing variety of forms, many quite bizarre, appeared along Ediacaran shores. Most disappeared, almost immediately, with no hint of progeny. By contrast, the descendants of just three or four others now dominate the planet.

Soon internal gas bladders began to add lift to soft gelatinous bodies and a multitude of 'floaters' rose from the depths to hunt in the richer, warmer pastures near the surface. Some even hoisted their gas bladders into the winds above, using them as sails, while they trailed their food gathering cells in the water below. The bluebottles that are so often swept on to Australian beaches are a relic of those times. Another such 'living fossil' is the sea pen. Its feather-like modern forms apparently found no need to improve on their Ediacaran design. Some live not far from Ediacara today, on the bed of the Spencer Gulf of South Australia.

FLINDERS RANGES, S.A.

This landscape (*BELOW*) originated as a broad, shallow, ocean basin, teeming with life, at the edge of the newly formed Australian continent. These curving ridge lines of scalloped hills are the remnants of folds that were pushed up when oceanic crust began to plunge beneath the edge of the continental raft. This process would eventually build the whole of Australia's eastern side.

Spriggina, FLINDERS RANGES, S.A.

Among the new forms to evolve were some bizarre experiments in body design. This three-lobed animal (LEFT) was one of them. No multicelled creature based on thirds has appeared since.

At the precise point when newer, more complex life forms, such as Spriggina (ABOVE) entered the fossil record, the earlier forms, such as Dickinsonia and some jellyfish, made their exit. One explanation is that among the new arrivals were the first multicelled predators and scavengers, and those without protection paid the price.

Tribrachidium, FLINDERS RANGES, S.A.

REPRODUCTIVE POLYPS, *Physalia sp.*, QLD.

Like modern bluebottles, many early Ediacaran creatures were little more than cell cooperatives rather than truly corporate individuals. But a little more than 500 million years ago, burrows in the seabed ooze announced the arrival of the well-muscled, fully-coordinated worm. Oxygen had reached about a tenth of the present level and the last few evolutionary barriers began to fall. Soon hard body deposits appeared, leading to external shells, internal skeletons, and finally, to a new kind of external armour known as chitin. This latter combination of proteins and minerals was much lighter and stronger than calcium-based shell, and chitin still protects the world's largest animal group, the arthropods. These include crustaceans, scorpions, millipedes, centipedes, spiders, mites and insects.

TRILOBITE FOSSIL, *Xystridura sp.*, Q.L.D.

PARVANCORINA

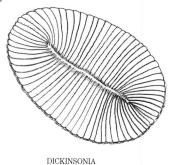

DICKINSONIA

RANGEA

The reproductive parts of the common Bluebottle, or Portuguese Man O'War, hang in a colonial cluster below its gas-filled float like a bunch of party balloons (LEFT). The colony includes both male and female polyps which cross fertilise and release embryos that grow into free-swimming larvae.

The most successful of all the early creatures were the trilobites. They launched a dynasty that was to last for 350 million years. Most were bottom dwellers and lived on continental shelves, so the distribution of their fossils often tells much about continental connections. Beetle Creek, in north-western Queensland, is so named because its crumbling banks abound with two species of them. One species, Xystridura (BELOW LEFT) is also found in Antarctica, North China and central Asia. The other, Lyriaspis (BELOW), is known from only one other region of the world, the Himalayas.

Among the earliest and most successful of the arthropods were a group known as trilobites, which diversified and spread throughout the world in a matter of a few million years. Superficially their fossilised remains resemble the modern garden slater, or wood-louse, though most of them were much larger. With chitin armour, a segmented, flexible body, and multitude of jointed legs, trilobites survived in a wide variety of forms for 350 million years.

Trilobites also seem to have pioneered one of evolution's most spectacular inventions, the compound eye. The same kind of complex mosaic construction has reappeared in two modern groups, insects and crustaceans, though neither can match the original version at its best. Some later trilobites evolved a method of uniting their chitin with crystals of calcium carbonate body-waste to form an array of excellently matched and highly efficient two-element lenses that were ideally suited to scanning dimly-lit ocean floors. A circular pattern of concentric ridging at the chitin-calcite interface preceded man's invention of the Fresnel lens by some 300 million years. [*Augustin Jean Fresnel showed that the volume of solid lenses could be significantly reduced by shaping them in concentric segments.*]

Though trilobites achieved wide distribution, most species remained relatively localised. In view of this, several early Australian trilobites display a pattern of family ties that is curious indeed. Close relatives now appear in fossil sites scattered throughout Asia. Family ties are strongest with fossils found in Malaysia, Burma, Thailand and China. There is even a group of trilobite fossils with Australian connections that have just been recently unearthed high in the mountains of western Tibet. It is from puzzling clues such as this that the story of Australia's past has been reconstructed. When a new piece of the jigsaw falls into place, it often explains a host of unrelated questions. The shape of Australia's trilobites and their Tibetan connection provides just such a piece.

TRILOBITE FOSSIL, *Lyriaspis sp.*, QLD.

600 – 500 million years

ʙARELY THREE HOURS *of our time scale remain. The land is still barren but warmth has returned to the tropics and marine life is flourishing. Oxygen levels are still climbing fast, despite an explosion of oxygen-breathing animal life in the equatorial seas.*

Australia, Antarctica and India form a nucleus of continents about which other scraps of young land crust, some still awash, are gathering in the southern hemisphere. Another continental cluster, including components of Africa and South America, is also pushing southward on the far side of the globe. As the two groups compact into a single, vast, 'ice-pack' of continents, shockwaves echo through their linkages as in the coupling of two trains. During this titanic jostling Australia's main internal seam gives way under the compression. It fractures and heaves into a chain of mountainous folds and overlaps that bisects the continent. This 2,000 kilometre birth scar is one of many in a supercontinent that will dominate the face of the planet for almost half a billion years. Its Australian scar will last even longer.

Formed from the debris of an intercontinental collision and weathered by heat, cold and innumerable storms, the Olgas are the result of a joining of continents.

THE OLGAS, N.T.

GONDWANA

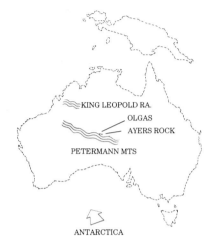

GONDWANA

If central Australia's Petermann Ranges (LEFT) give little indication of their original size, they give no hint of their significance. In fact they are the remnants of a mountain chain that erupted right across the continent when Australia was crushed by an intercontinental shock wave that was transmitted through Antarctica around 600 million years ago.

AT THE WESTERN END of Tibet, four of the world's greatest mountain chains meet. An awesome confusion of jagged, ice-bound peaks makes this one of the most spectacular and least hospitable places on Earth. The seasons deal harshly with all life here. Little moves and almost nothing grows except in the deepest valleys. Yet when winter loosens its grip on the mountains of Karakoram the frost-shattered boulders occasionally display impressions of animals that once lived in warm tropical seas. Even more remarkable, the genetic accent of some of them is distinctly Australian.

The explanation of their bizarre dispersal lies in the relentless drift of Earth's oceanic crust. These Karakoram fossils, especially the trilobites, seem to have originated in equatorial seas along Australia's northern coastline about 500 million years ago. This region consisted then of young and partly submerged land crust that later became fragmented and redistributed in the northern hemisphere. It now forms parts of central and South-East Asia, northern China, and perhaps even Siberia. Marine fossils with Australian connections have been unearthed in most of these former neighbours. Some of the similarities between their fossil faunas are slight, others are remarkably strong. The degree of resemblance is a good indication of how closely linked their habitats were at this time.

Australia's closest ties seem to have been with northern China, Tibet, and parts of South-East Asia. Some fossils found in Thailand and Malaysia precisely match species found in north-western Australia. Clearly these two regions were very close and probably contiguous.

This relationship is supported by certain features of Malaysian geology, and it is reinforced by magnetisation patterns fossilised in some South-East Asian rocks. There is evidence that Tibet's several components also formed a western extension of Australia at this time,

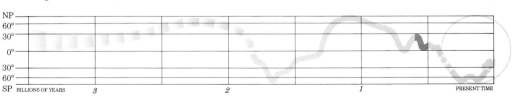

while beyond these lay a fragment that now contains the bleak Tarim Basin of central Asia. Little of these 'Asian' landmasses showed above sealevel, however, except part of the Tarim and North China blocks, and the whole region probably formed an archipelago.

This apron of emergent land crust along Australia's northern and western edges was part of a massive agglomeration of continents which bonded together in the southern hemisphere around 600 million years ago. The main components appear to have been South America, Africa, India, Antarctica and Australia. It was a momentous union. These five continents formed a supercontinental landmass that was to survive for almost half a billion years. We know it as Gondwana, a name drawn from a Sanskrit reference to an ancient Indian people; literally Land of the Gonds.

The shock waves from the bonding process seem to have echoed throughout Gondwana, showing mainly in the belts of thinner crust that formed the seams between older, more rigid components.

One of the shock waves reached Australia about 640 million years ago. This is recorded by rock deformities in the far south-west corner, though the main shock wave seems to have come a little later from further south. The effects of this were felt far inland, the evidence appearing mainly in central Australia. The shock reactivated the old central mobile belt all the way from the Canning Basin in north-western Australia to Broken Hill in the south-east. Over a period that may have lasted tens of millions of years this huge internal seam gave way once

The huge cluster of landmasses that formed the supercontinent of Gondwana (BELOW LEFT) appears to have come together between 650 million and 550 million years ago, with the final bonding process occurring between components of South America, Africa and Antarctica. It was probably a gigantic wave of compression produced by this last collision and 'echoing' through the continental linkages that ruptured Australia's central mobile belt to form the Petermann Mountains.

The debris washed from the original Petermann Mountains graduated in size from west to east. The Olgas are formed from boulder-strewn rubble, Ayers Rock from gravel and sand

The diagram (BELOW RIGHT) shows how later folding and weathering dictated the modern landforms.

THE PETERMANNS FROM MT. OLGA, N.T.

GONDWANA

TARIM BLOCK
INDO CHINA
KOREA
NORTH CHINA
S. CHINA
SE ASIAN BLOCKS
FRANCE
IRAN
SOUTH TIBET
SPAIN
AUSTRALIA
INDIA
AFRICA
MADAGASCAR
SRI LANKA
ANTARCTICA
NEW ZEALAND
FLORIDA
S. AMERICA

ADAPTED FROM BURRETT AND STAIT, 1987

more. Buckling and fracturing under the pressure, it gradually rose thousands of metres into the air, leaving the continent bisected by a 2,000-kilometre chain of snow-capped peaks.

Among geologists this is known as the Petermann Event, the name coming from a range of central Australian hills whose fold pattern is directly traceable to this compression. More evidence is printed in the rocks of the nearby Musgrave and Rawlinson Ranges. Worn to mere stubs, the modern Petermanns give little hint of their original size or significance, and most other direct evidence of the event has since eroded away or lies entombed in later sediments. But the original Petermann Mountains have memorials of another kind. As snow-fed mountain torrents began to eat into the chain of alpine peaks they flushed the debris into a remnant of the old central seaway lying along the mountains' northern flanks. Several slabs of this outwash material still lie in the emptiness of the vast Amadeus Basin. Now shrouded in heat haze, their sediments once more assume an aura of grandeur that does some small justice to their momentous origins.

Known to Australian Aborigines as Uluru and Katatjuta, two of these relics became the touchstones of tribal mysticism for thousand of years. As Ayers Rock and the Olgas they represent one of the meccas of modern tourism, though few know what they are looking at. Meanwhile, even as the Petermanns first began to rise some 600 million years ago, new patterns of convection turbulence were forming deep inside the planet. The general southerly drift of the Earth's crustal plates, which

(OVERLEAF): To the traditional owners of Katatjuta there were places among its many heads to be feared and avoided. Veiled and brooding in the dregs of a summer storm Wanambi-pidi (Mount Olga) well befits its awesome reputation as the lair of the legendary rainbow python.

(PAGES 106-7): Ayers Rock is best known for the remarkable procession of colours and shadows that daily pass across its facade, especially during sunrise and sunset. But just occasionally, when thunderstorms are brewing at dawn, an alchemy of light may push it to the frontiers of the surreal.

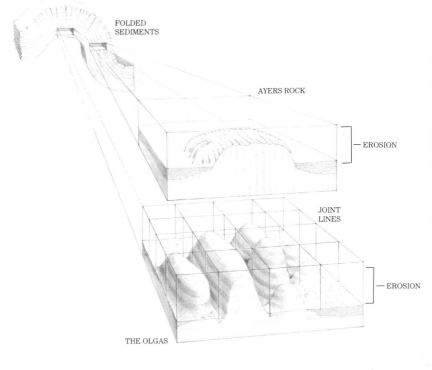

FOLDED SEDIMENTS
AYERS ROCK
EROSION
JOINT LINES
EROSION
THE OLGAS

Known as Ayers Rock and the Olgas these two outcrops are a mecca for tourists around the world: as Uluru and Katatjuta they were, for thousands of years, the repository of dreams for the Aboriginal people of central Australia: as remnants of debris from the original Petermann Mountains they commemorate the birth of Gondwana.

had gathered the continents in their polar cluster, now began to divide into an east-west pattern. The huge oceanic plate adjoining Gondwana's eastern side began to plunge beneath Australia and Antarctica along a 4,000 kilometre front. The friction and compression caused by this left its mark on Gondwana in the form of a mountain chain that ran unbroken for the full length of this eastern side. Though now in two parts this huge deformity remains spectacularly visible as the Transantarctic Mountains and their South Australian extension, the Flinders Ranges. From the air, the scalloped crests of the Flinders Ranges sweep northward in gently curving rows, like old skin pushed into wrinkles by an unseen hand.

During the next 200 million years this same process would result in the addition of the whole of Australia's eastern side as the continental raft accumulated scraps of new material from the underthrusting plate. By contrast, a new pattern of crustal rifts began to show simultaneously along Gondwana's northern fringes, carving away several of its 'Asian' attachments. North China and the Tarim block seem to have been the first to go, sometime between 500 million and 400 million years ago. These were followed closely by a north Tibetan fragment known as the Lhasa block.

This crustal movement, producing extensions on Australia's eastern side and fragmentation in the north, was the first sign that the crustal spread was migrating from the northern hemisphere to the south. The presence of the huge Gondwanan landmass was to have far reaching consequences on the biological explosion that was even then beginning along its northern coastlines. When the first plants and animals established their beachheads on Gondwanan shores, it offered a vast and varied hinterland into which they could disperse with maximum diversity. Meanwhile, as a cornerstone of the Gondwanan complex, Australia played a crucial role in the nurturing and shaping of life during this momentous evolutionary period.

AYERS ROCK AND OLGAS, N.T.

TIME IS RUNNING OUT. *Our clock shows that more than 21 hours of the time scale have elapsed, yet the land is barren. The shield of ozone gas accumulating in the stratosphere now glows a misty blue in the sunlight as the ozone's big molecules scatter both blue light and the dangerous radiations that lie beyond blue in the sun's energy spectrum.*

The colour of the Earth's skies is a signal that life may safely emerge from the water at last. Some filamentous algae are already well adapted to occasional exposure to air. These now give rise to forms which colonise the shoreline with the aid of rudimentary root cells and vascular stems that contain a ducting system for water. The first forms are leafless but all are tied to wetlands by their water-based reproductive process. Nevertheless they thrive, and their miniature forests entice animals to follow. The first of these are millipede-like vegetarians equipped with a multitude of legs and watertight 'space suits' of chitin that prevent them drying out. They have few competitors and no enemies, but this relative tranquillity is short-lived. As the 22nd hour of our clock draws to a close, an unmistakable warning is scratched into the sands of a West Australian beachhead: the hunters are coming ashore.

Lungfish have altered little in almost 400 million years. It was from a lobe-finned, air-breathing relative of early lungfish that the first four-legged land animals evolved.

LUNGFISH, *Neoceratodus forsteri*, QLD

INVADERS

MURCHISON GORGE, W.A.

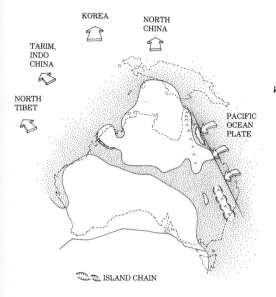

KOREA

NORTH CHINA

TARIM, INDO CHINA

NORTH TIBET

PACIFIC OCEAN PLATE

ISLAND CHAIN

INVADERS

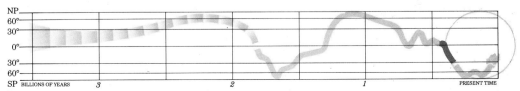
450 MILLION YEARS TO 350 MILLION YEARS (2123 HOURS – 2210 HOURS)

THE CHOCOLATE AND marzipan layercake of the massive Tumblagooda Sands makes the Murchison Gorge one of the most colourful in Australia. So well preserved are the 400 million-year-old ripples which pattern its floor that they might well have been left by yesterday's tide. But apart from worm tubes, the animal tracks etched into those ripples have no modern counterpart. They are the footprints of large marine predators known as Eurypterids, or sea scorpions, and they are among the oldest on Earth. The only other trackway of similar age is in Norway.

Armed with long forelimbs that were lined with spines and tipped by pincers, sea scorpions were at the head of the food chain and the terror of the seas at this time. Tracks up to 20 centimetres wide in Murchison Gorge suggest that the maximum length of these animals was a little less than a metre, though younger fossils twice that size have been found elsewhere.

Eurypterids seem to have separated from the basic arthropod stock almost 100 million years earlier. The Murchison tracks coincide with the movement of several groups into swamps, lagoons and river estuaries, where they became adapted to occasional foraging above the waterline. Ancestors of the modern scorpion emerged later, in much the same way. While the Eurypterid was doomed to extinction, this smaller, later cousin was to become one of evolution's spectacular successes.

But the very first colonisers of the land were millipede-like vegetarians and scavengers. They were closely followed by ancestors of the centipede and the spider, a little less than 400 million years ago and these were followed by insect-like scavengers, completing the food chain. The omen in the Murchison sands had been fulfilled: peaceful pasture had become hazardous jungle. And by then it was quite literally a jungle: evidence lies fossilised in the floor of an ancient Victorian

These ancient vertical worm tubes (LEFT) pattern the colourful sandstone cliffs at Kalbarri on the central west coast of Australia. Their original inhabitants lived in the delta sands that washed continually from the edge of the Yilgarn block, about 400 million years ago. As the sands accumulated above them, so the worms were forced to burrow upwards. Recently exposed by wave erosion, the sand-filled burrows proved more durable than the surrounding sediments.

The layered cliffs of Murchison Gorge (ABOVE RIGHT) are carved from the same sands. They appear to have fanned out over a broad rift valley that opened between Australia and what is now southern Tibet, when their bonds became stretched as northern Gondwana began to break up.

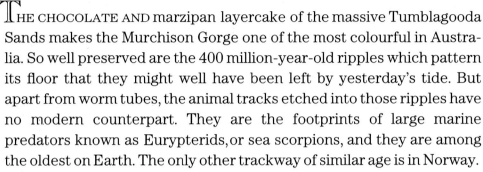

NP
60°
30°
0°
30°
60°
SP BILLIONS OF YEARS   3   2   1   PRESENT TIME

Looking a little like heavy-duty tyre marks, large animal trackways may be found in several of the sandstone horizons that now lie exposed on the floor of the Murchison Gorge, on Australia's central-west coast. Among the oldest footprints on earth, they were made by giant scorpion-like marine predators known as Eurypterids. These tracks, left on tidal sands about 400 million years ago, are the first sign of animals foraging above the waterline.

EURYPTERID TRACKS, MURCHISON GORGE, W.A.

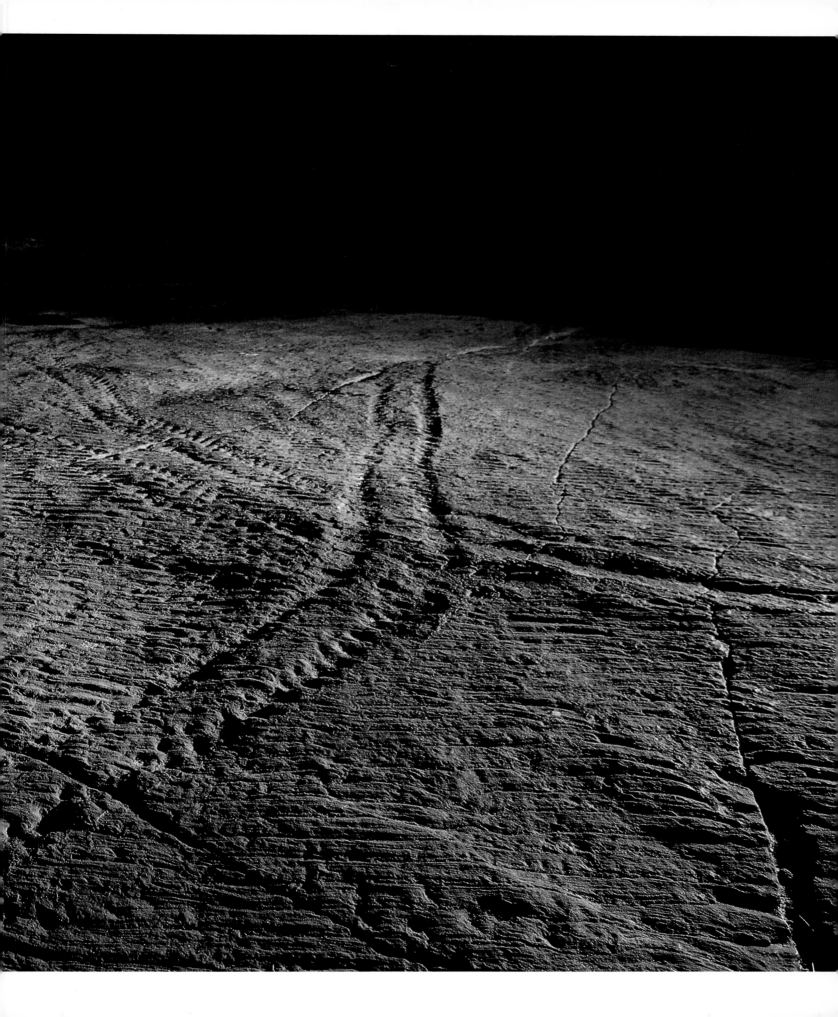

estuary where a frond from a clubmoss came to rest about 415 million years ago. It became impressed in the silt of the riverbed and the fossil today looks much like the shadow of a small feather boa. It is the earliest sign of a 'modern' plant in the fossil record and nearby fragments of marine fossils confirm its age. Within 30 million years, northern hemisphere relatives of that Victorian clubmoss were forming forests up to 45 metres high, while a host of new plant species vied for sky-light in the undergrowth. Among these were true mosses, liverworts, and the ancestors of a group which would one day dominate the world, the seed plants.

As primary producers, plants had been the natural pioneers. Though a few primitive leafless forms appeared in northern Africa some 430 million years ago, fragmentary evidence of fossil spores suggests that their history may go back a further 25 million years or more. Plant needs are simpler than those of animals, whose nature is inherently parasitic. They require merely water, sunlight, carbon dioxide and a few trace minerals. In colonising the land, the trick was to maintain a 'home' environment for their sea-bred cells. The walls of all marine plants were permeable to facilitate the passage of gas and soluble minerals. Their first adaptation for life out of the water was to cover their surface with a waxy skin, or cuticle, to preserve a watery environment within. Specialised breathing pores developed in this skin to regulate gas exchange. Direct food absorption also became impossible, so new ducting systems for the collection and redistribution of food and water had to be developed. These reticulation systems were reinforced by cellulose, which helped stiffen the stems that held the plants' photosynthetic surfaces up to the light. Spurred on by competition for light and a wider spore dispersal, herbaceous plants gradually evolved into trees. Larger plants needed more carbon dioxide to photosynthesise their food, a consequence of which was the creation of more oxygen.

The energy budget of the animal kingdom was similarly expanding. As muscle cells burnt oxygen at a prodigious rate to fuel movement that was no longer supported by water, so they increased the volume of carbon dioxide that they exhaled. With the waste gas of one providing the breath of life for the other, the exploding populations of plants and animals became totally interdependent, both chemically and physically.

Meanwhile, in the estuaries and lagoons of Gondwana many marine creatures were developing new physical attributes by depositing some of their body solids along a central body canal, thus forming a stiffened spinal cord to which muscles could be attached. The structure was proving its worth in a multitude of forms in fish. Protected only by bony

None of the earliest plant invaders survive in their original form, but six groups of their descendants have retained many of their most primitive features. Equipped with neither true leaves nor proper roots, five of these groups still thrive in Australia. Among the most graceful are the clubmosses, or Lycopods. Best known of these is the Queensland Tassel Fern, which grows as an epiphyte, high among the rainforest trees of north-eastern Queensland. Their hanging branches are each tipped with a 'tassel' of spore-bearing branchlets.

Most Lycopods look a bit like miniature conifers, and indeed, both male and female spores are born in cone-like structures. But unlike conifers they are restricted to permanently moist habitats by the requirements of their primitive reproductive system.

More primitive are the thalloid liverworts, a group of fleshy, ground-hugging plants that have no internal ducting system to distribute nutrients about the plant body. They also depend on the growth of a parasitic intermediate stage (toadstool-like structures) to lift their spore producing organs off the ground for better dispersal (RIGHT).

It is possible that one modern 'primitive' known as Psilotum may in fact be descended from one of the ancient plants it so resembles. Similar forms, known as Rhynia began colonising the world's wetlands more than 400 million years ago. Both rootless and leafless Psilotum now grows from moist cliff crevices and reproduces by means of spores released from capsules scattered along the main stem.

TASSEL FERN, *Lycopodium myrtifolium*, QLD.

MINIATURE CLUBMOSS, *Lycopodium cernuum*.

Psilotum nudum, SYDNEY OPERA HOUSE.

LIVERWORT, *Marchantia sp*, N.S.W.

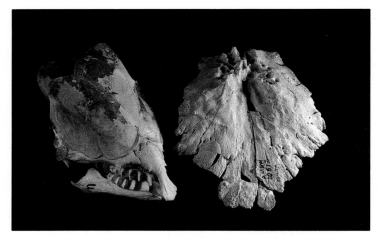

MODERN LUNGFISH SKULL, FOSSIL NOSEPLATE, QLD.

LUNGFISH, *Neoceratodus forsteri*, QLD.

370M YEARS OLD AND BREATHING

The ancestry of the human line may be traced back, with reasonable certainty, to a group of bony fish whose skeletons extended into their four, fleshy, lower fins. They diverged from other bony fish a little less than 400 million years ago. Later species of this line were also distinguished by a modified flotation bladder that enabled them to breathe out of water.

But far more graphic is the living fossil that still lurks in several of Australia's eastern rivers. This, the Queensland Lungfish, is a direct descendant and has changed little from its 370 million-year-old ancestors. The Australian species is the most primitive of the three surviving groups. It appears to be the oldest surviving genus – and species – of vertebrate in the world. The most pronounced change its family has undergone in the last 300 million years has been in size. The upper portion of a nose-plate from a 10 million-year-old specimen is as large as the entire skull of a modern lungfish. Its owner must have been some five metres long, four times the size of a large modern specimen.

head plates and a tough, scaly skin, these new creatures depended mainly on speed and manoeuvrability for survival.

The power source for these assets was already embedded in their chemistry since vertebrates had coopted an ancient bacterial innovation, the haemoglobin molecule, to serve as the oxygen carrier for the whole body. The haem component of haemoglobin is virtually a chlorophyll molecule, but with an iron atom at its core instead of a magnesium atom. This suggests that chlorophyll, the oxygen producer, evolved in an oxygen-free world, whereas haemoglobin evolved as a

Lungfish are weed-eaters. They lay their eggs among the reeds that line the banks of their rivers. Visible through the decaying membrane of its egg sac, is a two-week-old baby (BELOW).

LUNGFISH EGG, *Neoceratodus forsteri*, QLD

bacterial response to rising levels of oxygen pollution. In mammals, however, haemoglobin not only provides an oxygen delivery service for each cell, it also back-loads cell wastes, and thereby represents the lynchpin in a highly efficient respiratory system.

The success of this chemistry touched off a burst of vertebrate development. This is nowhere better displayed than in the sediments of a 380 million-year-old lagoon on the southern margin of the Kimberley block in Western Australia. The lagoon was shielded from open seas by reefs, part of an extensive tropical system that then fringed the

FOSSIL FISH, *Eastmanosteus sp.*, KIMBERLEYS, W.A.

The length and significance of the wall of jagged grey limestone that runs along the southern edge of the old Kimberley block only becomes apparent from the air. It is the remains of a barrier reef that formed around the Kimberleys when the region became semi-isolated by encroaching seas, some 370 million years ago. The lagoons that lay in its lee became a major nursery of early marine life, which the limey ooze of its floor preserved with an eerie fidelity.

Kimberley islands. Its outlines have been traced for hundreds of kilometres. Within the grey ramparts of this reef lies one of the world's richest deposits of ancient fish fossils.

Sinking gently to the floor in death, they were quickly buried and preserved within the limey ooze. Among the profusion of head plates, pieces of scaly skin, fins, and other bony fragments, are some in which the skeletal structure extends into four flipper-like lower fins. A descendant of one these lobe-finned fish would, within 20 million years, emerge from the sea and establish an entirely new dynasty of land life,

WEATHERED LIMESTONE, NAPIER RA, W.A.

BUNGLE BUNGLES, KIMBERLEYS, W.A.

BUNGLE BUNGLES, KIMBERLEYS, W.A.

BUNGLE BUNGLES, KIMBERLEYS, W.A.

The extraordinary array of sculptured domes and pinnacles that sprawl across the plains fringing the eastern Kimberleys seem to be a legacy of Gondwana's earliest dismemberment. Erosion has carved them from a thick sheet of layered sediments that washed into a rift valley there some 400 million years ago. The widespread rifting at this time seems certain to be linked to the tearing away from the Kimberleys of a broad apron of crust whose fragments later reassembled to form parts of South-East Asia.

the vertebrates. Somewhere among these fossils is our direct-line ancestor.

When these lobe-finned fish first began to move up rivers and into the swamps they were already well equipped for land life. Armed as they were with rudimentary 'limbs', powerful muscles and a tough, scaly skin, only one barrier remained to prevent their survival out of water. They needed a large gas exchange centre that could be kept moist, despite sun and wind.

By about 360 million years ago a solution had arisen. It consisted of a modified flotation bladder that was well supplied with a dense network of surface blood vessels. We know what it was like because it survives in Australia today, almost in its original form. The modern owner of this remarkable piece of equipment, the Queensland Lungfish, still resembles its historic forebears. It is one of only three species of lungfish left in the world. The other two, one each in Africa and South America, both have paired lungs and now depend wholly on surface breathing. In this they represent the next stage in the evolution of land vertebrates: when their rivers dry up during a drought they can burrow into the muddy floor, secrete a mucous cocoon and, breathing normally, survive in a state of torpor for long periods. Though the Queensland Lungfish never learned to combat this degree of desiccation—and it cannot burrow—it copes well with shallow or stagnating water by gulping air from the surface to supplement the oxygen breathed through its gills.

The lungfish's skeleton, being partly of cartilage, cannot support its weight on land. In water, however, it is able to use its fleshy fins as rudimentary limbs, propping itself on all four when resting on the river bottom. Lungfish still look much the same as they did 300 million years ago, though the immediate ancestor of the Queensland Lungfish grew much bigger, reaching a length of four metres in some cases.

The nearest living relative of modern lungfish is the two-metre-long Coelacanth, an equally primitive lobe-finned marine relic. It migrated to deep water about 70 million years ago and now finds refuge off the eastern edge of Africa's continental shelf. But it is a third and extinct member of this ancient family that means most to us, because with the aid of an additional air passage that linked its nostrils with the roof of its mouth, it established the beachhead for all backboned land animals. By 360 million years ago the first of the four-legged amphibians were making occasional forays along the river banks and tidal mud flats.

The oldest evidence for this too lies in Australia. In east Gippsland, not so very far from where the earliest clubmoss became fossilised, the

Some 480 million years ago the western Simpson Desert (BELOW) lay at the bottom of a warm, shallow sea, in which some of the world's first fish had begun to evolve. A scrap of that original seabed, containing its historic fossils, caps the hill in the background of this picture.

MOUNT WATT, SIMPSON DESERT, N.T.

(RIGHT): These tracks are among the earliest footprints of a four-legged land animal yet found anywhere. They were made by a creature rather like a modern newt that emerged from a river estuary on to a mud bank in eastern Victoria around 360 million years ago.

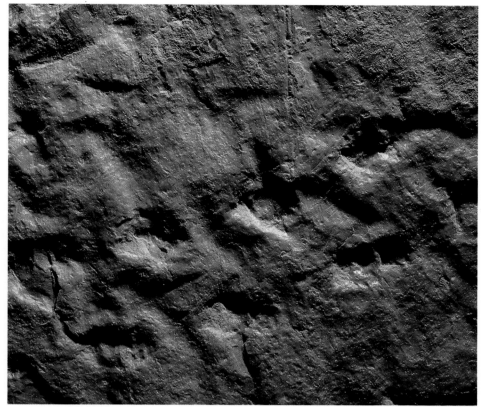

AMPHIBIAN TRACKS , GENOA R., VIC.

Genoa River has uncovered the remains of another ancient mudbank. One section of it, now housed in the National Museum of Victoria, bears two distinct sets of tracks. Printed there some 360 million years ago, they are the oldest tracks left by a four-legged land animal anywhere in the world. Some prints show signs of five toe impressions: and at one point in the tracks a shallow midline groove suggests that the body, tail, (or both) had been lowered on to the mud. With the aid of those four limbs, stout lungs, and a mouthful of sharp teeth, the genetic heirs of this cumbersome amphibian invader were destined to become lords of the Earth. Their collective name, labyrinthodont amphibians, is derived from the complex labyrinthine internal structure of their peg-like teeth.

Though they were awkward on land, these amphibians dispersed easily along suitable waterways. It was a timely talent. Earth's continents were about to cluster into a single gigantic mosaic, allowing amphibians to spread through all major continents. But around 330 million years ago mobility became vital for other reasons. Gondwana lay astride the south pole and, as snow clouds gathered once more in southern Africa, world temperatures began a fitful descent into another terrible ice age. For our amphibian ancestors it would be a case of the quick and the dead.

350 – 200 million years

THE WARNING COMES *comes at 10.20 p.m. on our clock: climates grow erratic, polar temperatures spiral ominously downwards and glaciers begin to gather in South American mountains. It is the birth of a new ice age. All the world's major landmasses are beginning to cluster on one side of the planet, gradually bonding to form a single landmass that sprawls almost from pole to pole. Coinciding with a general cycle of world cooling, this vast, brief conjunction, known as Pangaea, disrupts all globally rotating weather patterns. World temperatures plunge. Meanwhile, crustal drift carries Gondwana, the southern half of Pangaea, from one side of the pole to the other. The consequent migration of polar ice carves a 10,000 kilometre swathe of destruction across five continents, from South America to Australia. In its path whole mountain chains are ground to rubble and almost all life is extinguished. Nevertheless, hardship remains evolution's spur. Many new species appear and survival techniques, hammered out in the dry cold of those harsh winters, will preserve some of them for 300 million years. Two of these ice age innovations will become the pivots upon which most future evolution will hinge. They are the seed and the reptilian egg.*

These graceful 'wind sculptures' reflect turbidity patterns embedded in sediments by small eddies in the glacial meltwater that left them some 280 million years ago. Sheltered from the weather by an overhang, the more friable parts of the sandstone have since blown away.

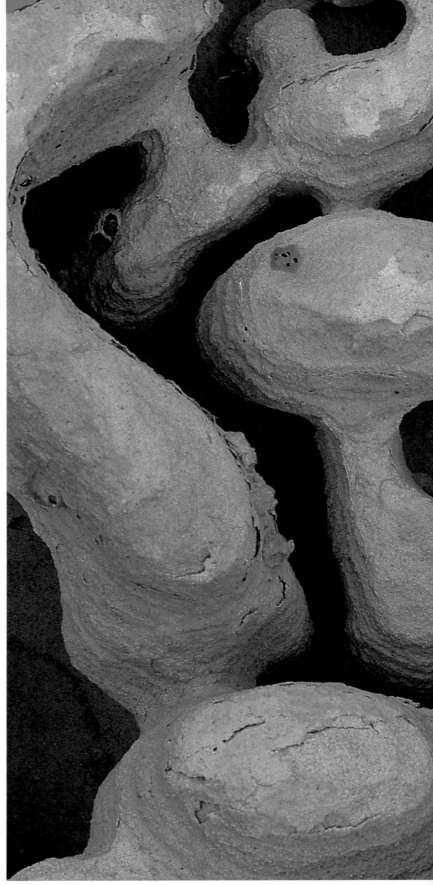

GLACIAL DEPOSIT, KIMBERLEYS, W.A.

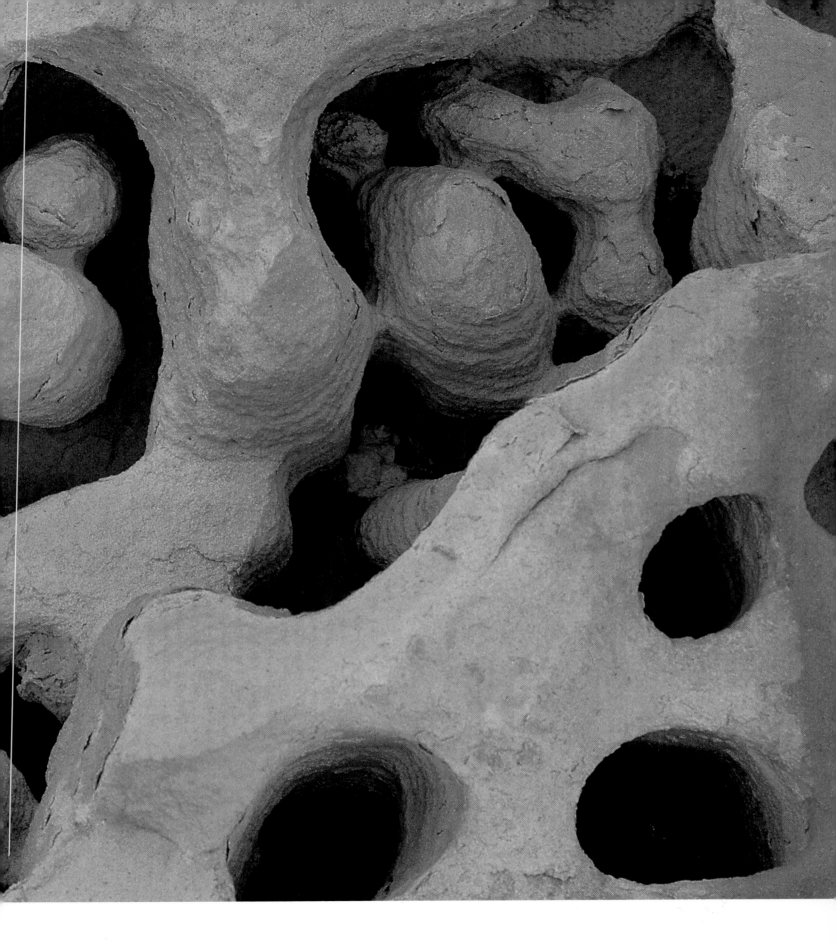

Spur of Adversity

SNOWDRIFT, MT. TWYNAM, N.S.W.

SPUR OF ADVERSITY

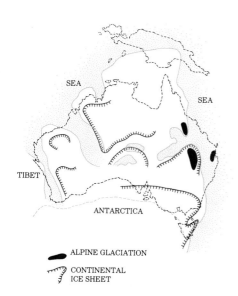

The clearest inscription a glacier can leave is the one that it cuts into its bedrock. One of the best examples of this (LEFT) stands on the cliffs overlooking Hallett Cove near Adelaide. The long grooves scored into the ice-polished surface of this glacial pavement were made by boulders embedded in the sole of the glacier that ground it flat. The gouge marks tell a detailed story of the glacier's passage including its speed and several changes of direction.

(ABOVE RIGHT): Grilled by the sun and undermined by meltwater, Australia's snowfields rarely linger in summer. This is the wind-sculptured underside of a summer snowdrift in the Snowy Mountains of southern New South Wales. The pinkish stains are due to a pigmented algae that often appears in late summer.

SNOWCLOUDS WHICH BEGAN TO gather in the mountains of South America and central Africa about 330 million years ago would not have seemed to be of any great significance to Australia. Lying on the opposite side of the huge Gondwanan landmass, Australia occupied approximately the same latitudes as it does today. But continental drift was about to bridge the 5,000 kilometre gap to the polar icesheet in just 30 million years.

Unlike the previous great ice age, this one only spanned about 50 million years and its main ice centres were concentrated around the poles. Its glacial surges were just as savage, however, and life had become more exposed and vulnerable because so much more of it was by then confined to the land.

Several factors probably contributed to the worldwide cooling, though the underlying causes remain uncertain. The curiously lopsided arrangement of the landmasses certainly played a major role. The north-south alignment of the combined continents disrupted all normal weather patterns and ocean currents, and the unbroken expanse of the single remaining ocean would have acted as a giant heat sink, reflecting little back to warm the atmosphere.

Large polar landmasses allowed ice to accumulate readily, and then continental drift compounded the effect in the south by spreading it along a 6,000 kilometre 'Polar Wander Path'. This apparent migration of the icesheet occurred as it maintained a static polar position, while the Gondwanan landmass drifted beneath it from one side of the pole to the other. Evidence of the destruction it caused leads from Bolivia to Queensland, and appears on five continents.

The driving force appears to have been an expanding ocean floor on the 'eastern' side of Australia and Antarctica. It had gradually herded

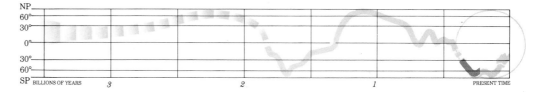

The sea cliffs on the north-eastern end of Maria Island off Tasmania's eastern coast, are built upon a thick sheet of glacial debris that fell from passing icebergs around 280 million years ago. Ice-polished boulders stud seabed sediments that are crowded with cold-water marine fossils. In some places you can see where colonies of living shellfish were crushed as boulders settled on top of them.

DROPSTONE, MARIA I., TAS.

As a tide of crustal drift carried Gondwana across the South Pole the ice sheets associated with the pole remained stationary, carving a swathe of destruction across the supercontinent that drifted beneath it.

The ice sheets that frequently entombed most of Australia during what is known as the Permian Ice Age left a trail of evidence that stretches more than 3,500 kilometres across the continent. Evidence of their passage is generally in one of two forms – rubble and rock powder dropped on seabeds from the undersides of glaciers, or thick blankets of unlayered sands that washed out from beneath them.

POLAR 'WANDER PATH'

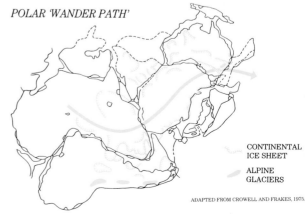

CONTINENTAL ICE SHEET

ALPINE GLACIERS

ADAPTED FROM CROWELL AND FRAKES, 1975.

132

GLACIAL OUTWASH, POOLE RA., W.A.

GLACIAL DEBRIS, FOSSIL BLUFF, TAS.

GLACIAL DEBRIS, KIMBERLEYS, W.A.

all the world's continents into a giant mosaic that stretched almost from pole-to-pole down one side of the globe. Most of Australia's eastern side was added during this process as scraps of new crust were plastered against the old continental 'shield', layer upon layer. The rugged basalt headlands around Port Macquarie were shaved from the surface of this oceanic plate as it plunged beneath the Australian continent and the lavas themselves were probably part of the convection cycle that pushed Australia under the ice cap.

Signs of the new ice age first appeared in Australia a little more than 300 million years ago. Glaciers began to gather in Tasmania and on coastal alps in what is now the Tamworth region of New South Wales. Snowfields then formed in the west, blanketing the highlands that represented the old continental foundations, the Kimberleys, the Pilbara and the big Yilgarn block.

A brief remission followed but it was the lull before the storm. Within a few million years a continental ice sheet began to grind northward out of Antarctica until it covered the whole of southern Australia. The icesheet's progress, and a later change of direction, is clearly etched into a fragment of its bedrock that now overlooks Hallett Cove, just south of Adelaide. Other evidence of its passage lies scattered right across Australia, from southern Queensland to the Kimberleys, with the bulk of it strewn along the old central seaway. In many places it lies above glacial debris deposited during Australia's previous glaciations, some 400 million years earlier.

At its peak the icesheet would have been at least four kilometres thick in places and the southern half of Australia sank more than half a kilometre under the weight. It was this huge downwarp that allowed the sea into central Australia again, even though sea levels were probably more than 200 metres below earlier levels.

Perhaps the most graphic evidence of the glacier's path is the ice-rafted debris of Tasmania. Here, along the north coast, where the main Antarctic icesheet first met the sea, the powdered rock of Antarctica's ground-down mountains now fans out from the foot of Fossil Bluff, near Wynyard, in the form of smooth, grey siltstone. Boulders of all kinds stud the surface, some deeply embedded, lying where they dropped from the underside of passing icebergs 285 million years ago. Recent erosion has removed the overlying rock to expose the original seafloor. A slightly younger boulder bed has been similarly uncovered along the shores of Maria Island on the central east coast. In some cases you can see where the boulders, carried by ice, have fallen on beds of living shellfish, crushing them.

Mud and sand washed from the base of melting glaciers now stamps its peculiar character on several Kimberley landscapes. These sandstones in the Poole Range have been 'filleted' by erosion, revealing convoluted turbidity patterns in the steady flow of meltwater that laid them.

GLACIAL DEPOSIT, KIMBERLEYS, W.A.

It was land life, however, that was most threatened and much of Australia's life was extinguished as glacial surges repeatedly engulfed most of the continent.

Despite the devastation, one group of survivors lingers still. These are small freshwater crustaceans known as syncarids. The largest of these, the Tasmanian Mountain Shrimp, achieved brief fame when first discovered because it proved to be a 'living fossil' in the truest sense.

One of a select group of animals that truly deserve to be called a living fossil is this small shrimp-like crustacean, the Tasmanian Mountain Shrimp. Its family evolved during the ice age, probably in the Australian-Antarctic region. By clinging to

MOUNTAIN SHRIMP, *Anaspides tasmaniae*, TAS.

cold, mountain environments, this genus has remained almost unaltered for nearly a quarter of a billion years. Its primitive features include external gills, branched legs and a fully segmented body, with no dorsal shield or carapace.

With its fully segmented body, exposed gills, branched legs and other primitive anatomical features, it is little changed from its ice-age ancestors. By clinging to mountain creeks and icy tarns, the environment which gave birth to its family, the mountain shrimp has survived practically unaltered for a quarter of a billion years.

Aided by their mobility in the waterways, Australia's labyrinthodont amphibians also survived. For them the thaw seems to have

released a tightly-wound genetic spring, and Australia became both a major stronghold and their last bastion. One 2.8 metre-long carnivore known as the Kolane amphibian, which lived in the lakes and waterways of eastern Australia, is the largest on record anywhere.

Strangely, frogs are all that we have left in Australia to remind us of this age of giant amphibians. It is the tadpole stage that most graphically betrays their origins. On the brink of adulthood, with four stubby legs and an eel-like tail, the humble tadpole resembles a miniature Kolane amphibian.

Among Australia's ancient plants, its tree ferns were perhaps the most notable survivors. They too have changed little and still flourish in cool southern forests.

On the far side of Gondwana, however, life had been dealt an additional wild card by the extraordinary conjunction of continents. While the north-south cluster of landmasses may have been the chief cause of the icesheets, it did offer room for manoeuvre to those lifeforms capable of exploiting it. As the deepening cold bit into life's fragile ecosystems, those species with a talent for dispersal survived well. And by the time the last continental glaciers had melted, land life had developed some remarkable new defences against such environmental stresses as cold and aridity.

The most serious handicap facing plants and animals in their colonisation of the land was the confinement of their reproductive cycles to water. It had shackled them to the wetlands and made them very vulnerable to drought. The problem had two facets: the male sperm of all plants and animals had to swim to its female counterpart for fertilisation to occur; the embryo then needed a constant waterbath to survive initial growth. Plants, the natural pioneers, first tried to bypass these requirements more than 300 million years ago, when some fern relatives encased their embryos in stoutly insulated, well-provisioned capsules. Though this group, known as seed ferns, was doomed to ultimate extinction, other seed plants soon began to evolve. They also introduced a waterless system of sperm delivery. The sperm of these

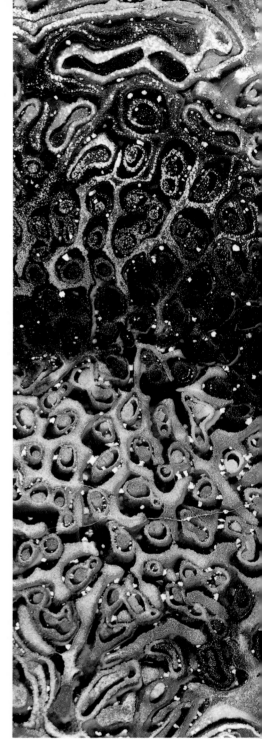

PETRIFIED TREE-FERN, TAS. (PHOTO: JIM FRAZIER)

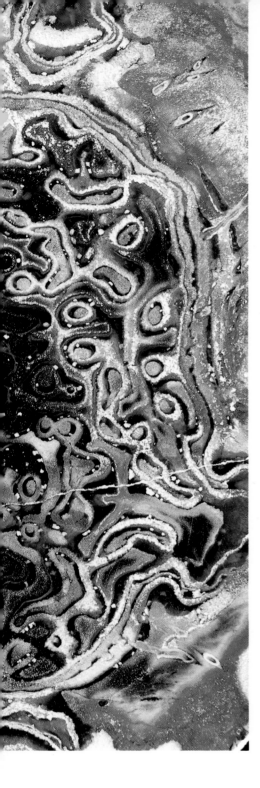

FERN FROND, *Angiopteris evecta*, QLD.

PRIMITIVE SURVIVOR THREATENED

This uncurling frond belongs to one of the world's most primitive tree-ferns, the Angiopteris. Its family was once widespread in Australia but they now appear to be on the verge of extinction. Only very small groups remain in the wild. Those shown here (ABOVE and OVERLEAF) are growing in a rainforest creek on Fraser Island, Queensland. (ABOVE LEFT): Fossilised stem sections of related tree-ferns have been found in several places on the mainland but none preserved as faithfully as the Tasmanian specimen shown here. This cross section closely resembles the peculiar stem structure of its relatives living on Fraser Island. There is no sign of a concentric growth pattern and no core of vascular tissue. The entire stem consists instead of evenly distributed vascular bundles.

The Cycad employs the most primitive system of seed-based reproduction in the world. It still depends on an external 'water' droplet to trap pollen; the male sperm carrier, and then the free-swimming sperm complete the fertilisation in an internal water chamber. Fertile female 'leaves' exude these droplets at egg sites along their stems: when pollen sticks to a droplet it is reabsorbed, fertilisation occurs and a seed begins to form (LEFT).

   Male cones (BELOW) shed large amounts of pollen to compensate for the vagaries of wind dispersal.

MALE CONE, Cycas sp, QLD.

FEMALE CONE AND SEEDS, Cycas media, QLD.

This skeleton of a modern rain-forest leaf (LEFT) hangs on the frond tips of one of the world's most primitive ferns, Angiopteris evecta. The pattern of veins in the skeleton leaf is characteristic of most flowering plants and contrasts with the ancient fern venation which has no cross-connections.

The net-vein pattern of Glossopterid leaves (BELOW) makes its earliest appearance in glacial sediments near Melbourne. Their ice-age success was due in part to the Glossopterids' habit of shedding all their leaves each autumn, a habit which contributed much to Australia's modern coal deposits and makes their leaves one of the commonest plant fossils.

GLOSSOPTERID LEAVES, N.S.W. (PHOTO: JIM FRAZIER).

Angiopteris evecta AND 'MODERN' LEAF, FRASER ISLAND, QLD.

plants was sheathed in sporopollenin, the most durable plant material of all. These armour-cased sperm parcels—pollen—could be delivered by the wind over long distances, right to the door of the egg chamber. Here, an artificial water droplet, exuded by the female, now awaited them—a waterbed tailormade for the consummation of the marriage. As soon as the egg was fertilized it was drawn back within the safety of the egg chamber where the resulting embryo could then be provisioned and stoutly packaged for extended survival on the ground, until the right conditions for growth unlocked the casing.

By the time the snowclouds began to roll away from southern Australia, some of these ancestral seed plants had already begun to disperse from the wetlands to dry, windy hillsides. They were the cone-bearing ancestors of modern cycads and conifers. Meanwhile, down in the swamps, a uniquely successful group of seed-fern descendants, known as glossopterids, was making a special contribution. Their leaves were broad and veined in a net pattern and in colder regions, such as Australia then was, they shed the lot at the onset of winter and grew a new set each spring. The broad working surface and the efficient veining of these leaves made the most of southern Gondwana's limited sunlight, while the tree's habit of winter hibernation reduced the risk of frost damage.

In a remarkable echo of the evolution of the seed, animals too achieved freedom from their ancient water bondage during this time. Somewhere in Pangaea, members of the amphibian family developed the modern system of intromission (male injection of sperm) and subsequently evolved a packaged embryo that could survive land birth. The elegant product is with us still — the well-provisioned, waterproof, air-breathing, reptile egg. In a modified form, we use this system in human reproduction. Armed with this egg-laying capability, a tough waterproof hide and, in some later cases, slightly warmed blood, these primitive reptiles launched a dynasty that was to last more than 160 million years. We know them as dinosaurs. For most of that time they dominated the land in forms that ranged from the size of a chicken to 75-tonne monsters.

Meanwhile, skulking in the shadow of the dinosaurs was another small, semi-reptilian offshoot. Members of this group eventually began to hatch their eggs internally and, with the aid of warm blood and an enlarged brain, they learned to live by their wits, foraging in the protection of the night. They dined on the dinosaurs' leftovers, perhaps a reptile egg or two, and bided their time. Their descendants, the mammals, would not emerge from the shadows until the dinosaurs had left the scene.

200 – 130 million years

W HEN THE LAST HOUR *begins, the Earth is*
on the threshold of dynamic change as a living
planet. Most of its surface is already inhabited
by a multitude of forms that are complex and
well adapted to their environments.
Their day is slightly shorter than ours, the
climate is warmer and there is little or no snow,
even at the poles. Despite these differences we
would instantly recognise the planet as our
galactic home. The ancient conifers now
dominate the hillsides, with the first of the
modern cycads scattered amongst them. Ferns
and treeferns clothe the valleys. Few of the
creatures which roam these forests have much
in common with modern forms. Some
amphibians remain but in many parts of the
world reptiles have taken over: the evolution of
the reptile egg has helped to launch the world's
most successful animal dynasty. The first
dinosaurs enter Australia around 11.00 p.m. in
our time scale.
But even as the early forms of dinosaurs
disperse through Pangaea, it tears in half and
begins to fragment. Convection currents deep
inside the planet are reaching out to stir the
biological stew.

The crumbling pillars of dolerite that characterise so
much of Tasmania's mountain regions are the geological
equivalent of scar tissue, a continental 'stretch mark'
acquired early in the launching of this southern ark.

BEN LOMOND, TAS.

STIRRING THE STEW

SALAMANDER FISH, *Lepidogalaxias salamandroides*, W.A.

STIRRING THE STEW

Iɴ ǫᴜɪᴇᴛ ᴄʀᴇᴇᴋs that flow from the great karri forests of south-west Australia lurks a small eel-like fish. It has permanently flushed cheeks, soulful eyes, and an expression of deep disappointment. It has no close relatives in the whole of the southern hemisphere. A faint structural resemblance can be seen in the large and voracious pike family that inhabits the cold freshwaters of Europe and North America. But there it ends. The Australian fish, *Lepidogalaxias salamandroides*, is much more primitive than the pike in many of its features. It can also breathe air and burrow into its creek bed to escape drought, a talent that has earned it the name Salamander Fish.

Some 2,500 kilometres to the east another lonely relic lives the hermit life in cold, wet Tasmanian caves and the dank recesses of rotting logs. This is the elegant—and very large—Cave Spider, *Hickmania troglodytes*. Its leg span may reach 170 millimetres. Like the Salamander Fish, this too is the sole Australian survivor of its branch of the family, though members of two related families are scattered across the world, in Chile, Argentina, the USA and China.

Extinctions have clearly played a major role in isolating both of these evolutionary relics, but the Cave Spider is especially interesting. None of its relatives disperse aerially as do the spiderlings of many modern species, and they all depend on the availability of cold, damp hiding places. In cases like this, where all members of the group are relatively immobile or easily stopped by environmental barriers such as water or desert, such a distribution used to be baffling until it was realised that the continents themselves move, rafting their inhabitants about the face of the planet. Only recently has it become known quite how far and how fast continents travel. It is now arguable that Pangaea represented a mid-cycle cluster in a regular pole-to-pole migration of the world's landmasses.

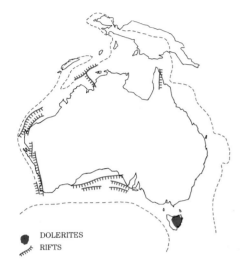

● DOLERITES
〜 RIFTS

The far flung family connections of many Australian plants and animals were puzzling until it was realised that the continents themselves had moved about the planet, rafting their inhabitants to their present positions. Locked to a habitat of moist darkness in caves and hollow rainforest logs, Tasmania's Cave Spider (LEFT) is a relatively immobile species. It has been isolated from its nearest relatives by the drift of continents.

(ABOVE RIGHT): This small fish, shown emerging from its summer hiding place in the sands of a creekbed, is a relic as old as the Tasmanian Cave Spider. Its only relatives appear to be a northern family. An ability to live out of water, breathing air, has earned it the name Salamander Fish.

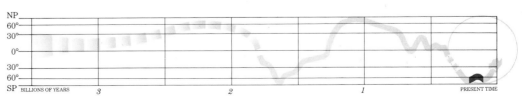

NP
60°
30°
0°
30°
60°
SP   BILLIONS OF YEARS   3   2   1   PRESENT TIME

CAVE SPIDER, *Hickmania troglodytes*, TAS.

Like the Tasmanian Cave Spider, the West Australian Salamander Fish is a relic from a time when most of the world's landmasses were linked together in a single supercontinent known as Pangaea. When Pangaea fragmented, the extinction of other southern members of its widespread family meant that its only living relatives were restricted to the north. This species now lives in the sandy creeks that flow out of the Karri forests of southwestern Australia. Among its survival techniques is the ability to burrow into its creekbed during drought and breathe air until the creek fills once more.

SALAMANDER FISH, *Lepidogalaxias salamandroides*, W.A.

For plants and animals, however, the Pangaean conjunction and subsequent fragmentation could hardly have come at a better time. Many widespread Pangaean families became dispersed around the world as a result of the inexorable drift of their continental homelands. In this fashion convection currents deep inside the Earth appear to have been even more important than other dispersal mechanisms in spinning the web of life on this planet.

First signs of Pangaea's breakup appeared in its equatorial waistline, roughly where the Mediterranean and the Gulf of Mexico now lie. The rift that opened there almost 200 million years ago effectively divided all land life into two broad evolutionary streams, one centred in the northern continents and the other in the great southland, Gondwana. This basic evolutionary divergence underpins the 'Australian' character of many of our so-called 'primitive' plants and animals.

It was during this time too that extensive rifts began to appear along Australia's western coastline, marking the departure of southern Tibet. The final scar, a massive straight-line fault known as the Darling scarp, parallels the present coastline for a thousand kilometres. To the west of it the Australian remnant of the old weld line slumped below sea level. The existing coastal plain was built on this slumped section. Only two small fragments of the old weld line remain visible, one embedded in the coastline near Northampton and the other, semi-detached, forming the blunt 'chin' in the south-west corner of the continent.

Shortly after southern Tibet broke away, signs of stress between Australia and Antarctica began to appear in Tasmania. The evidence remains in the massive pillars of dolerite that stand in Wagnerian grandeur on Tasmanian mountaintops and coastal cliffs. The dolerite was originally injected as molten lava into huge subterranean fractures that opened there about 175 million years ago, when the intercontinental bonds first came under stress. Cooling more quickly than the main mass of molten rock, the upper surface of the dolerite sheet solidified first, developing a regular fracture pattern due to shrinkage.

The columnar dolerite that characterises so much of Tasmania originated far underground, when lakes of molten rock filled huge subterranean fractures that appeared between Australia and Antarctica some 175 million years ago. These fissures were the first signs of the crustal stresses that would ultimately tear the billion-year-old bond apart and set Australia free. Where they are now exposed to weathering, the edges of the dolerite slabs form 'organ pipe' structures. (BELOW, RIGHT)

THE ACROPOLIS, DU CANE RA., TAS.

This island, known as Sugarloaf Rock (LEFT), and the nearby coastline form one of the few tangible legacies of Australia's long association with what is now Tibet and the Indian subcontinent. This section of coast, between Cape Leeuwin and Cape Naturaliste, represents a fragment of the weld line that neither properly detached nor submerged.

CAPE NATURALISTE, W.A.

BEN LOMOND, TAS.

The most dramatic colonnades of Tasmanian dolerite are those exposed in the sea cliffs of the south-east and on the ice-carved mountain tops in the west. Best known is Cape Raoul (ABOVE), a crumbling palisade that thrusts into the Southern Ocean from the tip of the Tasman Peninsula, not far from Hobart. It is one of many such headlands on a coastline which is largely carved from dolerite alone.

Locked in the mountain wilderness on the western side of the island a similarly dramatic colonnade crowns a feature of the Du Cane Range. It is known as the Acropolis and echoes, in its decay, the majestic architecture of classical Greece (RIGHT).

CAPE RAOUL, TASMAN PENINSULA, TAS.

THE ACROPOLIS, DU CANE RA., TAS.

These cracks gradually extended downward to the base of the sheet. Where the dolerite is now exposed to weathering, the hairline fractures gradually split open, the columns separate, and one by one they fall away.

During the tectonic fragmentation, first of Pangaea and then Gondwana, the breaking of the land links between the continents became crucial in the ultimate distribution patterns of both plants and animals. Where closely related species occur on continents that are now widely separated, it is a good indication that the modern sea

155

FOSSIL: *Agathis jurassica* TWIGS: WOLLEMI PINE, *Wollemia nobilis*, N.S.W. (PHOTO: JAIME PLAZA)

A *tiny relictual population of Wollemi pines (LEFT) growing in a secluded mountain gorge near Sydney was recognised from the peculiar nature of the litter on the forest floor. Unable to shed individual leaves these pines shed each lateral branch after it has borne a cone at its tip. The yellowish female cones (visible near the crown of the tree on the right) generally grow above the rust-coloured male cones, limiting self pollination.*

(ABOVE RIGHT): A small twig from a Wollemi pine, laid beside leaf impressions left by a long-extinct common ancestor of both the Kauri and the Wollemi pines, shows how little the Wollemi has diverged from its ancestral stock in 150 million years.

barriers had not opened when their ancestors dispersed. One of the more successful dispersals between 180 million and 150 million years ago was that of the conifers. Riding the wind by pollen and seed, two ancient southern families, the Araucarians and Podocarps, gradually established themselves throughout Gondwana.

Both families are still well represented along Australia's eastern coastal fringe, although the Araucarians now dominate in the north, while most Podocarps have adapted to cooler southern regions. Some of these, notably the Huon and Celery-top pines, along with the delicate Miniature and Creeping pines of the alpine heath, are now wholly confined to cool, moist Tasmanian habitats, whereas Araucarians, such as the stately Bunya, Hoop, Kauri and Norfolk Island pines, no longer grow in Tasmanian forests at all.

Between 94 million and 30 million years ago another member of the Araucaria family was also common throughout the Australian–Antarctic region. We know this because its pollen is abundant in the fossil record and has even been found in off-shore drill cores. Sadly, no living descendants remained – or so it was thought. In 1994 however, one small population of descendants was discovered in a little-known canyon in the Blue Mountains north-west of Sydney. The entire population consists of no more than 40 adult trees and their seedlings.

Known as the Wollemi pine, its present status as one of the world's rarest trees is a direct consequence of Australia's long voyage north from Antarctica and the onset of the current ice age. These factors produced a much dryer, more seasonal climate, a change that proved intolerable for the Wollemi pine which gradually disappeared from all of its former habitats – except one.

STIRRING THE STEW

Podocarps were among the first conifers to make their appearance in Australia. Among the best known of their descendants are Tasmania's Huon pines (BELOW). Like most conifers living in cold climates their rate of growth is very slow, and although they do not achieve great height, the growth rings of some logged specimens show that they may live two thousand years or more.

This miniature relative of the Huon, the Creeping Pine (BELOW RIGHT), clings close to the ground among the cushion plants of the wind-blasted alpine heath.

HUON PINE *Lagarstrobos franklinii*, TAS.

CREEPING PINE *Microcachrys tetragona*,   CUSHION PLANT, *Dracophyllum minimum*, TAS.

PENCIL PINE *Athrotaxis cupressoides*, TAS.

Shaped by the rigours of their alpine habitat, Pencil Pines may not grow large, but they can survive for hundreds of years. This ice-scarred specimen (ABOVE) grows on a glacial shelf on the eastern flank of Mount Mawson, in Mount Field National Park.

PENCIL PINES, *Athrotaxis cupressoides*, TAS.

CELERY-TOP PINE, *Phyllocladus aspleniifolius*, TAS.

Australia's most unusual conifer, the Celery-top Pine, is a member of one of the most ancient pine families, the Podocarps. Betraying little of their conifer origins the 'leaves' of this young Celery-top (LEFT) are not true leaves, but phyllodes – flattened stem material that is well supplied with chlorophyll and breathing pores, and serves the same purpose.

By 130 million years ago the conifer invasion of Gondwana was complete. The next wave of invaders, the flowering plants, had not yet emerged from their equatorial birthplace in western Gondwana and would not reach Australia for almost 30 million years. But in those cool, fern-filled southern forests, where the last of the world's labyrinthodont amphibians were fighting a losing battle against their ancient adversaries, the crocodilians and the dinosaurs, an even greater revolution was beginning. Old rifts along Australia's southern margins were stirring into life. Gondwana was about to tear apart at the seams, and the new invader would be the sea.

130 – 65 million years

Aᴛ 11.20 P.M. *on our time scale a comet bursts out of the southern skies and plunges to Earth in the very centre of Australia. The explosion shakes the planet and leaves Australia with an umbilical scar 22 kilometres across. Gondwana has already begun to tear apart, and marine flooding is widespread because of exceptionally high sea levels. But, for the moment at least, life thrives. Both Australia and Antarctica are covered in forest, mainly conifers, cycads and tree ferns. These in turn support a large dinosaur population and an underworld of small mammals, including the ancestors of both marsupials and monotremes. It is into these luxuriant forests that evolution launches one of its most elegant weapons, the flower. Plants have learned that by decorating and baiting their genitalia they can lure animals into service as pollen couriers. In this way they achieve pinpoint accuracy in their pollen delivery, and the sweet smell of their success spreads swiftly round the world.*

This is the standard blemish on the face of an ageing planet, a cosmic impact scar. It was made by a wandering fragment of star debris that plunged to earth in central Australia about 130 million years ago.
The original crater, much larger than this, has long since eroded away.

GOSSE BLUFF, NT.

AN OMEN

CAVE, GOSSE BLUFF, N.T.

AN OMEN

LAWN HILL

WINTON

GOSSE BLUFF

L. ACRAMAN

CRUSTAL RIFT

CAPE OTWAY

ANTARCTICA

◼ LAND AREA
↓ DINOSAUR TRACKS
• MAJOR IMPACT SCARS

The layers of hard sandstone that form the inner walls of the Gosse Bluff amphitheatre (LEFT) originally lay between two and three kilometres below ground level. They erupted to their present position during the gigantic rebound explosion that followed the impact.

(ABOVE RIGHT): Dawn gilds a cave at the foot of the inner walls.

NEAR THE VERY CENTRE of Australia stands a curious circle of hills. They rise abruptly from the flat semi-desert that surrounds them, enclosing a rounded amphitheatre some two kilometres across. Its walls are formed by an up-turned collar of hard sandstone that rises almost vertically to a crown of jagged peaks. It is as though the plain had once bubbled and then burst.

This is Tnorula, or Gosse Bluff.

The whole planet was once pock-marked with such blemishes in various stages of decay—the Moon still is— because Gosse Bluff is one of the trademarks of an evolving universe. A piece of cosmic debris, probably a comet, crashed to earth there about 130 million years ago.

It would have come in as a silent fireball, outrunning its shockwave, at speeds up to 30 kilometres a second. Plunging into the plains near the western end of the Macdonnell Range, it penetrated little more than half a kilometre before vapourising. It released something in the order of 10^{20} joules of energy in the process. Measured against the grim scale of human achievement in this field, it produced more energy than the combined output of 200,000 Hiroshima-sized atomic bombs. The shockwave pulverised a huge hemisphere of the underlying rock and sent earth tremors racing round the planet.

The rebound explosion that followed this compression then disembowelled the plain immediately beneath the vapourisation chamber. Almost 400 square kilometres of Central Australia erupted into the air. Some rock, now at the surface, came from more than three kilometres below, and slabs up to half a kilometre long were flung into the air. A mushroom-shaped cloud would have punched almost 20 kilometres into the atmosphere and quickly spread round the world, staining southern hemisphere skies for months afterwards.

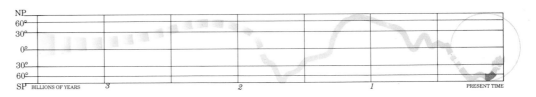

NP
60°
30°
0°
30°
60°
SP BILLIONS OF YEARS   3   2   1   PRESENT TIME

AN OMEN

THINGS WERE DIFFERENT...

Australia lay near the South Pole when the Gosse Bluff comet arrived. The Southern Cross and other stars recorded here were not then in the positions that they now occupy, and the shattered rock in the foreground lay undisturbed almost four kilometres below the Earth's surface — beneath lush, green plains and the realm of dinosaurs.

Radiating fracture lines, such as these (RIGHT) are characteristic of high-velocity cosmic impacts. Where such fractured rock becomes exposed to weathering it falls apart along these fracture lines to reveal their cone pattern.

SOUTH POLAR STARFIELD, GOSSE BLUFF, N.T.

SHATTERCONE, GOSSE BLUFF, NT.

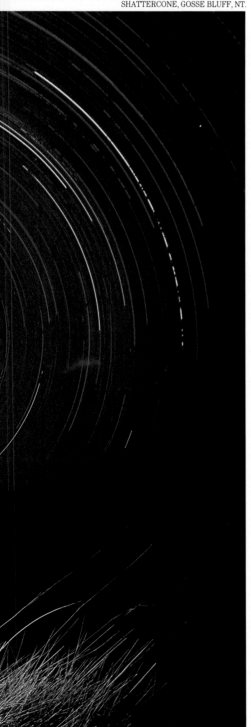

Gosse Bluff is drained by a single creek which has carved a winding gulley through the eastern wall (BELOW). It allows easy access to an area that originally lay directly below the comet's vapourisation chamber. Rock now exposed at the centre of this amphitheatre originally lay more than three kilometres underground.

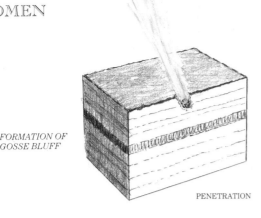

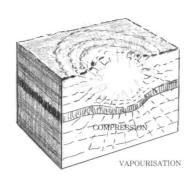

*FORMATION OF
GOSSE BLUFF*

PENETRATION

COMPRESSION

VAPOURISATION

DECOMPRESSION

REBOUND

GOSSE BLUFF FROM MT. PYROCLAST, N.T.

The circle of hills that we know as Gosse Bluff is merely the core of this rebound explosion. The rock layers that form the walls of the central amphitheatre have been identified, horizontal and undisturbed, two kilometres below the surface of the plain surrounding Gosse Bluff. When the dust first cleared, this fractured collar would have been embedded in the floor of the crater and buried beneath half a kilometre of debris. The crater itself was some 22 kilometres across and perhaps a kilometre or more deep from its rim. All this overburden, including almost all trace of the original crater, has since been removed by erosion.

A characteristic signature of such high-velocity impacts is the radiating fracture pattern which fragments all rock in the immediate area. When exposed to weathering the rocks tend to fall apart along these cone-shaped fracture lines, hence their name, shattercones.

A second major impact zone has been recently detected in Australia because of the presence of these shattercones. Known as the Lawn Hill Structure, it is about 250 kilometres north of Mount Isa in western Queensland. It was made by a slightly smaller missile more than 70 million years earlier; and almost nothing of the scar remains visible on the surface. Underground, however, its symptoms are precisely the same as those which showed in seismic surveys of Gosse Bluff. Both areas have a huge underlying bowl of shattered rock. If this rubble could be replaced in its original position, all of its shattercones would point to the source of the initial shockwave.

A third impact scar, on the Eyre Peninsula in South Australia, is far older and less discernible—but of a scale that dwarfs the recent ones. The site, almost undetectable at ground level, was first noticed during scrutiny of satellite photographs. The clue was a set of very faint concentric circles, up to 190 kilometres in diameter, centred on a large salt lake. These circles proved to be ancient fracture lines but their cause remained untraceable. The riddle was solved when the fracture zone was correlated with a layer of curiously structured conglomerate rock 300 kilometres to the east, in the Flinders Ranges. The salt lake

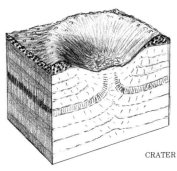

CRATER

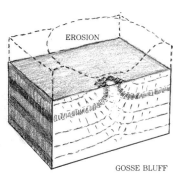

EROSION

GOSSE BLUFF

The Gosse Bluff explosion hurled huge slabs of rock astonishing distances. Mount Pyroclast, five kilometres to the south, is the remains of one such slab of debris. It affords the only good view of the southern walls of the Bluff (BELOW). Here too, the rock is shattered in the same cone-like pattern.

ALLOSAURID FOOTPRINT, C. OTWAY, VIC.

Australian dinosaurs are of peculiar interest in that they represent the southernmost limit of dinosaur dispersal and the only example of polar dinosaurs yet found. With a footfall that in some cases could be measured in tonnes, they left their unmistakable signature at several sites which then lay within the Antarctic Circle. One of these is from Cape Otway, Victoria (ABOVE). The foot that compressed the original mud thereby endowed it with a durability that eventually preserved it when most of the surrounding material weathered away.

At Lark Quarry, near Winton in central Queensland, however, it is the original impressions that remain – more than 4,000 of them, excellently preserved. The 'flow pattern' of of the trackways and many other clues suggest that more than 150 herbivores, most of them small, had been put to flight by the arrival of a giant meat eater, or carnosaur (RIGHT).

The large print (ABOVE RIGHT) was not part of the stampede however, but had been made a little earlier by a big plant eater.

FOOTPRINT, *Wintonopus sp.*, LARK QUARRY, QLD.

concealed a gigantic impact scar and the layer of conglomerate, containing jagged lumps of Gawler Range lava, was the fallout from the vast dustcloud.

If we want to express the energy released during this Lake Acraman impact in the same terms as those used for Gosse Bluff, and still keep the figures comprehendable, then we must up-grade the basic unit of explosive from the 'tiny' Hiroshima A-bomb to the largest weapon ever tested. This was an H-bomb with the destructive power of 100 million tonnes of TNT. Even using bombs of that magnitude it

TRACKWAY, CARNOSAUR (large), LARK QUARRY, QLD.

Rhoetosaurus brownei

AN ALLOSAURID

WHEN DINOSAURS DOMINATED...

DRAWINGS: PETER SCHOUTEN, *'Prehistoric Animals of Australia'*

Judging by footprints found in southern Queensland, the first dinosaurs appeared in Australia almost 200 million years ago. Among the oldest skeletal remains are those of a 17-metre-long plant eater that died between 170 million and 180 million years ago. Unearthed near Roma, Queensland, this appears to have been a distinctly Australian species of lizard-hipped dinosaur, *Rhoetosaurus brownei*. It stood about three metres high at the hip, but with its long neck erect, its head would have reached almost double that height. It may have weighed up to 20 tonnes. Moreover, *Rhoetosaurus* was not the largest. A solitary neckbone found near Hughenden, Queensland, appears from both its shape and size to have come from a close relative of a giant African form, which reached lengths of 20 metres or more.

Most of the larger dinosaurs were plant eaters, browsing on Australia's pines, cycads, and treeferns, but here and there among early dinosaur tracks are footprints which belong to a group of carnivores resembling the north American *Allosaurus*. Up to ten metres long and four metres tall, these dinosaurs generally carried their massive weight on their hind legs alone, leaving their forelimbs free to grasp prey.

Precisely where dinosaurs originated remains uncertain. However, despite Australia's poor fossil record, most groups seem to have appeared here as early as they did elsewhere. But once protected by Australia's isolation, early forms probably survived here longer. As a consequence, Australian dinosaurs frequently developed characteristics that were particularly regional.

The most complete dinosaur skeleton ever found in Australia was that of a seven-metre-long, bird-hipped dinosaur, discovered near the Queensland township of Muttaburra. *Muttaburrasaurus langdoni* was a semi-bipedal with long hind legs and bird-like feet. It had an elongated snout that was tipped with what appeared to have been a horny beak, and though it had no teeth in the front part of its mouth, the curious construction of its cheek teeth suggested that it may have added a little meat to its basically vegetarian diet. It had lived around 120 million years ago.

Many marine reptiles moved inland with the seas that inundated much of the continent between 100 million and 110 million years ago. Among them was the world's largest aquatic predator of those times, *Kronosaurus queenslandicus.* Its skull alone was almost four metres long, more than a quarter of its total body length, and it was armed with a formidable array of long, conical teeth that interlocked like those of a modern crocodile.

Also in those seas were two ancient reptile groups that would survive the extinctions that were soon to swallow the entire dinosaur dynasty. These were the ancestors of our turtles and crocodiles.

Muttaburrasaurus langdoni

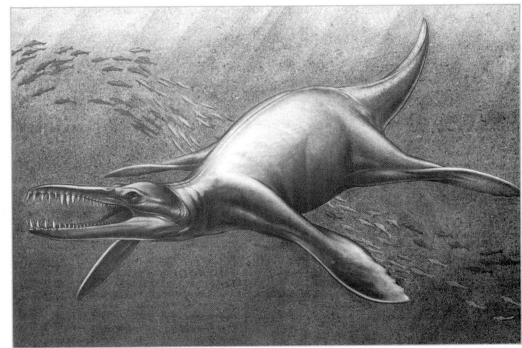

Kronosaurus queenslandicus

RELATIVE SIZE TO MAN

ESTUARINE CROCODILE, *Crocodylus porosus*, N.T.

would require the simultaneous detonation of 50,000 to 100,000 of them to match the explosion that occurred at the Lake Acraman site a little more than 600 million years ago.

By tradition comets have been typecast as the harbingers of plagues, famine and the fall of kings. There is now some evidence that they may in fact have given life a head start on Earth by acting as preliminary incubators of organic chemistry while the planets were being formed. Each time the Earth then passed through the tail dust of a comet it would have been 'infected' by their complex organic

molecules. This might well explain the 'indecent haste' with which life first appeared on Earth. The recent Australian comets were at least two billion years too late to add to it however, for the oxygen which helped to fuel their fiery end would also have destroyed any life-building potential in their tail dust.

The Gosse Bluff explosion was the first in a series of major southern hemisphere events. Within a few million years, India had detached from Gondwana, almost half of Australia was submerged, and the last of the labyrinthodont amphibians had become extinct. On the land, as in the sea, reptiles now reigned unchallenged.

One of the best illustrations of this has been uncovered near Winton in central Queensland. There, on lakeside mudflats 100 million years ago, a giant predator surprised a group of bipedal dinosaurs, most of which were little larger than chickens. In the stampede that followed they left more than 4,000 footprints. It is one of the biggest and best-preserved trackways sites in the world.

Another important find of this age was made just 250 kilometres to the west, near Boulia. Bones collected there confirmed that flying reptiles known as pterosaurs, had once existed in Australia. This is the only evidence of the pterosaurs yet found in Australia, though they had appeared in the northern hemisphere some 90 million years earlier. They soared around the world on bat-like membrane wings. All the more remarkable then are the delicate impressions of seven bird feathers, some 20 million years older, which were discovered in freshwater sediments near Koonwarra in south-eastern Victoria. Among the oldest feather fossils in the world, they were from warm-blooded descendants of small predatory dinosaurs.

Such early bird forms would have been less vulnerable in flight than their membranous-winged predecessors. Feathers enabled them to more readily alter the effective area and shape of their wings so that take-offs, landings, and sudden wind gusts were rendered less hazardous. A feathered wing was also less easily damaged and, in most instances, readily repaired because feathers were regularly replaced.

A *fraction of a second before this picture was taken the water was absolutely calm, its surface reflections broken only by the crocodile's nostrils and eyes. The sheer speed and efficiency of their attack is one good reason why they remain unchallenged as the dominant predators in their various habitats.*

Like the pterosaurs, birds nevertheless retained their ancestors' teeth at this stage. The last toothed birds seem to have died out as late as 60 million years ago. It was at this time also that the plant world launched its most beautiful weapon, the flower. By enlarging, decorating and baiting their sex organs, plants enticed animals, especially insects, to visit them frequently during their most fertile period so that they might dust them with pollen. The flowers could thus use their visitors as unwitting sexual couriers.

This reproduction method enabled flowers to proliferate much more economically than plants which depended upon the vagaries of the wind to deliver their pollen. The flower represented an outlay of a little leaf material, a little protein pigment, sugars, water and (in some cases) a dab of perfume to provide a long-range guidance system for the pollinator. This got their pollen delivered to another member of the species with pinpoint accuracy. By comparison, those plants that depended on the wind to carry their pollen, such as pines and cycads, had to produce vast quantities of valuable genetic material to achieve anything like the same fertilisation rate.

The oldest flower pollen recorded in Australia belongs to an ancestral member of the ancient holly family. This worldwide group now has only one Australian representative, *Ilex arnhemensis*, a tree which grows in a few isolated pockets in monsoonal areas. Other species occur in New Guinea and New Caledonia. But fossil *Ilex* pollen has been found right across the continent, especially in the south, and the existence of a species in New Caledonia suggests that they must have entered the region before the opening of the Tasman Sea, some 80 million years ago.

The first flowers evolved between 120 million and 130 million years ago, almost certainly in the tropics and probably in either Africa or South America. From there they gradually spread around the world, reaching southern Gondwana some 10 million to 15 million years later. They moved into Australia in increasing numbers during the next 40

Ilex arnhemensis, N.T.

The very first flowers to appear in the Australian region included some that looked a little like this. They belonged to a representative of the Holly family which arrived here more than 100 million years ago when much of the continent was waterlogged and half of it lay below sea level (LEFT).

ANTARCTIC BEECH, *Nothofagus moorei*, QLD.

Australia's floral link with South America shows clearly in the distribution of the rainforest trees known as southern beeches (ABOVE). Among the earliest of the flowering plants, their South American ancestors spread through the southern continents some 90 million years ago. They now not only occur in Australia and Tasmania, but in New Zealand, New Caledonia and the highlands of New Guinea. Significantly they do not show up in the fossil record of either Africa or India. It suggests that the ocean barriers to these two continents were substantial by this time.

A small mammal jawbone containing three teeth that became preserved in the potch of a New South Wales opal field has proved to be one of the most momentous fossil finds of recent times. It came from an animal that appears to have been the common ancestor of both the platypus and the echidna, the world's only egg-laying mammals, or monotremes.

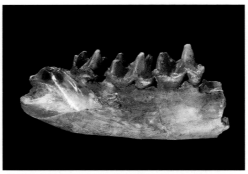

MONOTREME JAWBONE, AUST. MUSEUM. (PHOTO: JOHN FIELDS)

This ancestral form appears to have evolved in the Australian region from a very early offshoot of the mammal line, and lived in the wetlands of eastern Australia during a period of very high sea levels and widespread inundation a little more than 100 million years ago. No sign of monotremes has been found outside Australia.

million years, supplying the genetic rootstock for much of our modern flora.

Among the most successful of the first flowering plants were the *Nothofagus* or southern beeches. Taking advantage of the much milder climates that prevailed in polar regions some 80 million years ago, they dispersed easily through the temperate forests of southern Gondwana. Although there are only three Australian descendants in the cool, montane forests of Tasmania and the mainland, others are scattered around the southern hemisphere in South America, New Zealand, New Guinea and New Caledonia. Typifying what is now accepted as a standard Gondwanan dispersal pattern, *Nothofagus* distribution provided the crucial evidence that won wide acceptance for the theory of continental drift.

Travelling with these plant invaders was an army of small, well-adapted forest animals. Predominant among them were birds, many of whom had already been seduced into their modern roles as pollinators by the plants' gaudy flowers and copious gifts of nectar.

Marsupials too, were on the move, having evolved in the Americas, probably in the north, a little more than 100 million years ago judging by the fossil evidence. Marsupials feature strongly in South America's fossil record of 70 million years ago. Their teeth and bone fragments have also been found in Antarctica, yet no fossils older than about 54 million years have been unearthed here in Australia. On the other hand we know that New Zealand's crustal components had begun to detach from Australia by 80 million years ago and the absence of marsupials there suggests that they had not yet crossed Antarctica. So their arrival here remains a mystery.

Even more tantalising, however, is the riddle posed by Australia's two families of monotremes. The Platypus is the sole survivor of one of them and the other consists of two species of Echidnas: one widespread in Australia, the other, an older form, now confined to the highlands of New Guinea.

ECHIDNA, *Tachyglossus aculeatus*, S.A.

PLATYPUS, *Ornithorhynchus anatinus*, QLD.

While their bone structure and egg-laying habit suggests that as a group they diverged very early from the mammalian line, they are now highly evolved and finely tuned to their specialised environments. But there is little clue as to where or when this curious group originated. A beautifully preserved, opalised jaw and its teeth, found at Lightning Ridge in northern New South Wales, appears to be from a common ancestor of the Platypus and Echidna and has been dated at more than 100 million years old. (Modern adult monotremes have no teeth although they are still born with them.) A solitary monotreme

PIONEER VALLEY, EUNGELLA NAT. PK., QLD.

tooth found recently in Patagonia suggests that they must have radiated westward from their evolutionary home in Australia and achieved a wide distribution in Gondwana around 60 million years ago. The highly specialised nature of the two living genera also makes it likely that this ancestral group must have been very large, diverse and successful, having fully occupied the Australian–Antarctic region before wilting under the impact of the marsupial invasion.

By contrast, placental mammals failed to establish themselves in Australia before Antarctic ice closed the last migration corridor.

The gradual rise of Australia's eastern highlands was due to the opening of the Tasman and Coral Seas, and provided a vital mountain retreat for a multitude of hard-pressed plant and animal families during the later onset of ice-age aridity.

The distribution of species is often governed by the opening and closing of such physical barriers. Australian species that originated between 100 and 80 million years ago could still disperse through New Zealand to New Caledonia, and through Antarctica to South America, but they usually failed to conquer the widening seaways that then separated them from Africa and India. After the Tasman sea began to open about 80 million years ago, however, only Australia and South America remained linked to Antarctica. When matched against this continental fragmentation pattern, the distribution patterns of modern plants and animals provide a guide to where and when they originated.

In many cases flowering plants took their pollinators with them when they dispersed. This was especially true of the plants that had cultivated mammals and birds as pollinators (rather than insects) because these animals required specialised flower construction for the exchange of pollen to work well. These peculiar modifications for mammals and birds lie at the heart of most 'typically Australian' flowers, such as dryandras, banksias, hakeas and grevilleas.

With a surfeit of honey and pollen offered by such 'custom designed' flowers, their bird and mammal pollinators thrived. This ancient bondage remains strong among Australia's modern birds, but later aridity, which slashed Australia's forests to their present shreds, reduced the mammal pollinators to one—the Honey Possum of south-western Australia.

In this respect the Honey Possum, and a few other species, such as the Marsupial Mole, the Tasmanian Devil, the Numbat and the Bilby, represent our closest links to those resourceful marsupial ancestors that traversed Antarctica and founded a dynasty among Australia's monotremes between 55 and 65 million years ago.

Sometime between 65 million and 66 million years ago a series of extinctions began throughout the world. It occurred in all kinds of environments, in the sea as well as on the land, and reverberated down through time in ways that are not yet properly understood. What is certain, however, is that by the time the chain reaction came to a halt, many plants and almost a quarter of the world's animal families had vanished. But among rotting reptilian carcasses there was still the occasional scurry in the gloom. Living on their well-honed wits, the mammals had somehow survived. It had been a long apprenticeship but when the dinosaurs' mantle of supremacy fell to these unlikely successors they began to evolve with astonishing speed. And among them, in America, was a single primate, Purgatorius.

AS OUR STOPWATCH *begins the last twenty minutes of the time scale, temperatures take their first serious plunge since the previous ice age. Vegetation everywhere undergoes massive change and a series of extinctions begins to sweep through the animal kingdom. The most dramatic of these is the disappearance of the dinosaurs. But with the reptile dynasty in tatters, the mammals are free at last to emerge from the shadows.*

Meanwhile, Antarctica is being lassoed by a series of crustal rifts which tug it south once more, and its billion-year-old bond with Australia begins to give way. With 18 minutes to go, a 3,000-kilometre-long rift appears between them. As it widens its margins sag and the sea floods in from the west, leaving the two land-masses briefly linked by only a narrow isthmus which runs through Tasmania. This last exit closes with about 16 minutes to go, as the isthmus slumps into the sea.

Antarctica is locked at the pole and with another ice age looming, its cargo of land life is doomed. But Australia has escaped. Intact and well stocked, the great southern ark gathers way and heads north with its live cargo, alone.

Australia's last land link with Antarctica sank into the sea somewhere south of Tasmania about 53 million years ago. The collapse of this lingering corridor to Antarctica marked the end of a continental marriage that had survived for a billion years.

POINT VIVIAN, TAS.

LAST EXIT

PORONGORUP RANGE, W.A.

LAST EXIT

IN THE SOUTH-WEST CORNER of Australia, two rows of hills, the Stirlings and the Porongorups, lie side by side about 30 kilometres apart. They are neither large nor particularly spectacular but the events that they commemorate are monumental. The Porongorups are all that remain of a huge reservoir of molten granite that formed far underground during Australia's traumatic collision with Antarctica more than a billion years ago. From the Porongorups the Stirling Ranges are usually visible as a shadowy silhouette lying along the northern horizon. Their jagged height is a product of the more recent rupture of that same intercontinental bond. By their origins these two ranges thereby mark both the beginning and the end of Gondwana, the supercontinent that fostered so many of the Earth's ancestral lifeforms.

The rise of the Stirlings was punctuated by another event of even more spectacular proportions. A series of massive changes were about to transform both plant and animal kingdoms. They began about 65 million years ago as extinction after extinction cascaded down through the food chains, wiping out a quarter of the world's animal families, including the dinosaurs, and an unknown number of plants. Though the bulk of the extinctions seem to have occurred almost simultaneously in geological terms, they probably sprawled across hundreds of thousands of years, perhaps even a million or more.

The best-founded explanation proposes that the catastrophe was triggered by the impact of a comet, probably one that had already fragmented to some extent. Such a collision, or perhaps a series of them, would have thrown up a dust cloud so large that it blanketed the entire planet, filtering out the Sun's light and warmth to the point where much of the world's plant and animal life perished. The plants would have died first, causing animal food chains to collapse.

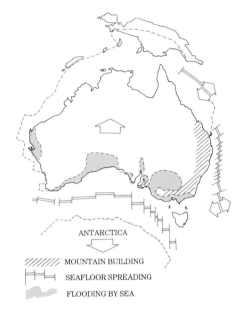

ANTARCTICA

////// MOUNTAIN BUILDING

SEAFLOOR SPREADING

FLOODING BY SEA

The seabeds that formed the Stirling Ranges in south-western Australia began to rise to their present height within the last 100 million years. They now form the largest mountain range in the southern half of the State. This view from Bluff Knoll, the highest point, looks towards Pyungoorup Peak, which is wreathed in cloud, at the eastern end (LEFT).

Formed from molten rock during the bonding of Antarctica and Australia, more than a billion years ago, the ancient granites of the Porongorups lay parallel and barely 30 kilometres to the south of the Stirling Ranges. The two ranges thereby commemorate both the birth and death of the super-continent of Gondwana, of which Australia and Antarctica were cornerstones (ABOVE).

The entire Stirling Range seems to have squeezed out like toothpaste from a tube during the separation of Australia and Antarctica. The cause probably lay in the slow, uneven tearing process. During the early stages of their separation, the rift opened gradually from east to west as the two continents pivoted slightly about a western hingepoint. The rotation was small but probably sufficient to pinch the Stirling's sediments against the basement of ancient Yilgarn block immediately to the north, and squeeze them high into the air.

STIRLING RANGE, W.A.

The strongest evidence for this scenario appears in a layer of peculiarly contaminated dust that settled around the globe, mainly in the northern hemisphere, about 65 million years ago. The dust contained traces of the very heavy metal iridium, in a rare isotopic form. Iridium was present in the original material that formed the solar system and remains in comets, planets, meteorites and asteroids. However, most of the Earth's iridium was drawn into its molten core by fractionation during its early development. Only the arrival of fresh solar material from outside could have injected so much iridium-charged dust into the atmosphere. Even more important, the micro-structure and chemistry of this dust layer precisely matches the fallout from the Acraman impact site, thereby corroborating the comet theory.

A chain of impact craters of similar age has since been identified, girdling half the globe and ending in Central America. The huge crustal bruise at the tip of the Yucatan Peninsula is so large that the explosion that created it would have entirely emptied the Gulf of Mexico. Up to 75 per cent of the world's plant and animal species were extinguished in the aftermath of that unimaginable impact. The evolutionary door had been flung wide open, and first through it were the mammals.

Having spent 130 million years as bit players, scuffling in the darkness at the edge of a dinosaur-dominated stage, these small, shrew-like creatures now had the stage to themselves. With their warm blood, enlarged brain and long experience of living on their wits they were well prepared for the challenge, and the speed with which they began to diversify was dazzling indeed. The coincidental break-up of Gondwana also allowed isolated populations to evolve with maximum variation into the huge ecological vacuum left by the dinosaurs. Australia's very diverse marsupial fauna appears to be the product of precisely this scenario.

Antarctica was caught inside a noose of oceanic rifts at this time and was being gently tugged back to the pole. As soon as Antarctica had

This crumbling colonnade of sea cliffs overlooking Bass Strait (RIGHT) was originally formed from lava injected into huge subterranean fractures that opened there about 60 million years ago. The fractures were a symptom of the massive stress caused by the tearing apart of the billion-year-old bond between Australia and Antarctica and the stretching of Bass Strait.

The brooding cliffs of Precipitous Bluff (BELOW LEFT) look south over the stormy Southern Ocean where Antarctica once lay. This southern face became exposed by erosion after Tasmania's last land links with Antarctica slumped into the sea, between 50 and 60 million years ago.

PRECIPITOUS BLUFF, TAS.

The jagged cliff line that defines the southern edge of the Nullarbor Plain (LEFT) echoes massive submarine escarpments that lie far to the south, on the floor of the Great Australian Bight. The original escarpments formed as the edges of the Australian continental raft, slumped into the sea during the opening of the Southern Ocean between Australia and Antarctica.

TWELVE APOSTLES, VIC.

In western Victoria the big swells of the stormy Southern Ocean have carved a seabed of similar age and origins into one of Australia's most dramatic coastlines. Its towering cliffs and rock stacks (ABOVE) now form the centrepiece of Port Campbell National Park.

torn free from Australia's southern edge, the sea flooded in from the west to inundate the country's sagging continental margins. These inundations included much of the south-west coast, most of the Nullarbor region, the Murray River basin, Bass Strait and several small basins in Victoria. As the rift grew wider, Australia rotated slightly and compression in the south-west corner pinched the continental shelf against the ancient Yilgarn block. Wedged between parallel fractures, a long slab of seabed gradually became squeezed upward to form the Stirling Range.

By 55 million years ago the break with Antarctica was almost complete. Australia's only surviving land link lay through Tasmania. Bass Strait was at least partially submerged but Tasmania, in two offset halves, still clung to the main Australian plate. Tasmania's western half lay against Antarctica's Pennell Coast and, as Australia drifted northward, this provided a sliding contact that offered Antarctic lifeforms an escape route that lasted long after other land bridges had submerged. Antarctica still retained a tenuous link with South America through the Falkland Islands ridge, although this probably consisted of little more than a chain of islands.

Antarctica's last land link with Australia slid into the sea south of Tasmania about 53 million years ago. Minor biological exchanges continued while the water gaps were still small but about 40 million years ago Australia became reattached to the fast-moving Indian Ocean plate and was accelerated northward. As the polar rift pattern extended, a deep-water channel formed around Antarctica and the modern circumpolar ocean current developed. A sympathetic weather pattern then locked in cold polar air and Antarctica's fate was sealed. A polar icecap was about to close over the whole continent, extinguishing almost all its land life. Meanwhile, loaded with Gondwanan lifeforms, Australia headed for the tropics. The southern ark had escaped.

———

53 – 20 million years

Sixteen minutes *remain on our time scale.*
Life on Earth has almost recovered from one of
the most devastating series of extinctions in its
history. Birds and mammals now dominate.
World climates remain unstable, however, and
temperatures begin the long unsteady descent
that signals yet another ice age.

As a polar icecap begins to close over Antarctica
once more, Australia, the great southern ark,
drifts slowly northward towards the tropics.
The continent emerges from cloudy southern
latitudes into temperate sunlight, and dry
winds begin to unravel its thick cloak of
vegetation. Plants and animals are forced to
look to their defences: they must adjust now,
not only to a deteriorating world climate but to
a shift in latitude of some 25 degrees.

The immediate enemy is desiccation. The
options are to adapt or migrate. Many species
do become refugees, finding shelter in river
gorges, on cloudy mountain tops or along more
humid coastlines. But the typical 'Australian'
character will be forged by those that hold their
ground, modifying their structures and their
lifestyles to keep pace with the changing
conditions. Construction details for
specialisations as disparate as the toughened
gum leaf and the kangaroo's long middle toe
are hammered out here.

The Honey Possum may be the last of its line but its family
almost certainly played a major role as pollinators in
developing Australia's present plant diversity.

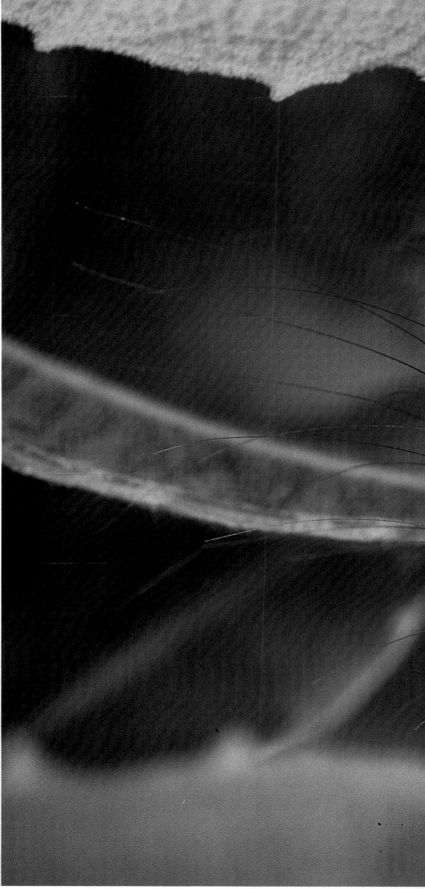

HONEY POSSUM, *Tarsipes rostratus,* W.A.

GREAT SOUTHERN ARK

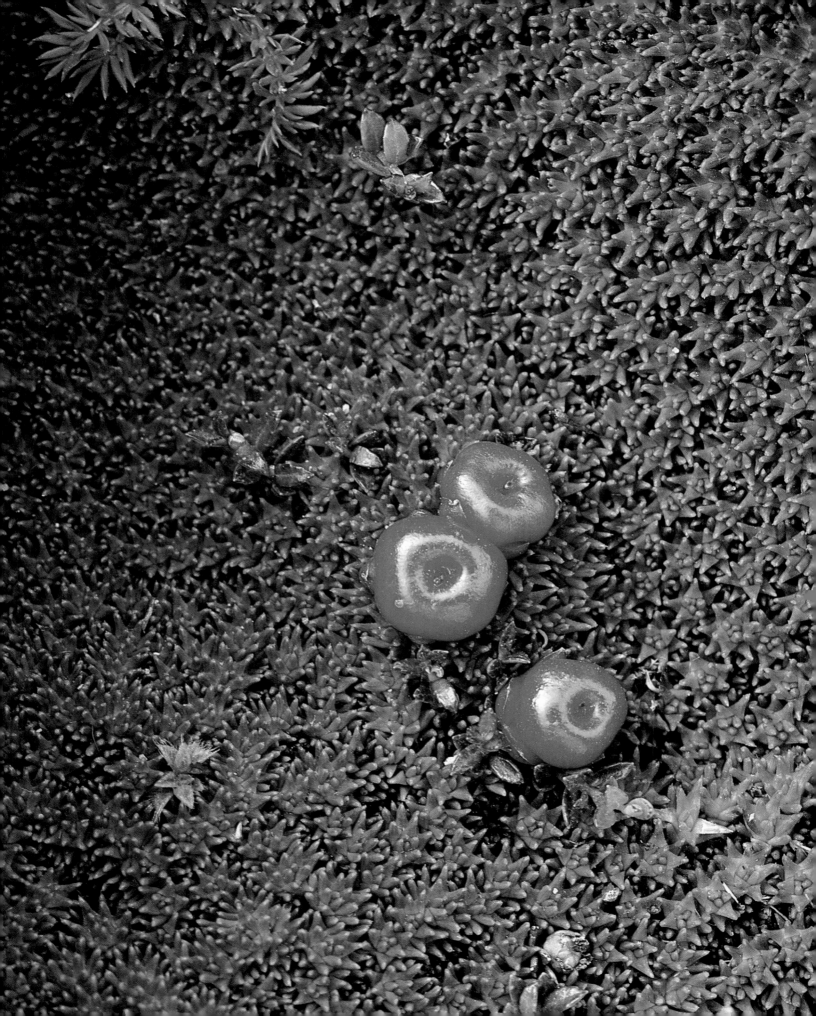

LEATHERWOOD, *Eucryphia lucida*, TAS.

GREAT SOUTHERN ARK

Aₗₘₒₛₜ ₐₗₗ ₒf ₜₕₑ plants and animals that we think of as characteristically Australian originated in the old southland, Gondwana. They represent a cross-section of ancient genetic material that was sealed into a continental 'test-tube' for 30 million years. It was subjected to heat, cold and aridity, and then gradually re-exposed to the genetics of the outside world via an island corridor. Australia offers no motley group of 'living fossils' but a spectrum of highly-evolved modern life, developed for the most part in total isolation. Such life affords us a unique insight into the mechanisms of evolution.

When Australia's land bridge to Antarctica sank into the sea about 53 million years ago only a few floating seeds, some birds, flying insects and other airborne migrants such as minute eggs, spores and spiderlings would reach Australia during its long voyage north. When a circumpolar current closed around Antarctica about 35 million years ago, temperatures dropped sharply. Soon tongues of ice began licking at the last of the forests that had once cloaked the continent.

The growth of polar icecaps is usually accompanied by a drying out of the Earth's mid-latitudes, so when Australia pushed northward out of the cloud cover of the Roaring Forties, its northern forests began to wither. Rainforest survived only in the cooler south and along the well-watered east coast.

Yet careful analysis of family distributions shows that many Gondwanan life forms remain scattered through tropical Australia, even in the far north, despite massive invasions by vigorous South-east Asian species. Along the east coast the old southern families now tend to dominate the high ground because they are better adapted to its colder, poorer soils.

CORAL SEA RIFT

ANTARCTICA

SHIELD VOLCANO
PATH TRACED BY MANTLE 'HOTSPOT'

Wedged among the tightly packed stems of other alpine cushion plants, this berry-bearing relative of a major African family (LEFT), the Ericaceae, grows only on cold Tasmanian mountain tops. Its restrictive habitat severely limits its ability to disperse, and seems to cut explanations for this isolated population down to one – continental drift. Even more intriguing, one of its relatives is the rhododendron, whose ancestral home is in the Himalayas.

Delicately perfumed Leatherwood blossom (ABOVE) owes its genetic origins to South American ancestors. On both continents the family is still confined to cold mountain rainforests. This Tasmanian species is responsible for the aromatic qualities of much forest honey during it's flowering season.

NP
60°
30°
0°
30°
60°
SP BILLIONS OF YEARS   3   2   1   PRESENT TIME

Delicate as a bonsai garden, this mat of highly specialised cushion plants clings to a wind-blasted ridge on one of Tasmania's highest mountains. But its genetic links reach half way round the world. While its family home is in south-western Australia – with a major branch in India – this particular genus is shared only by Tasmania, New Zealand and South America.

CUSHION PLANTS, *Phyllachne colensoi*, TAS.

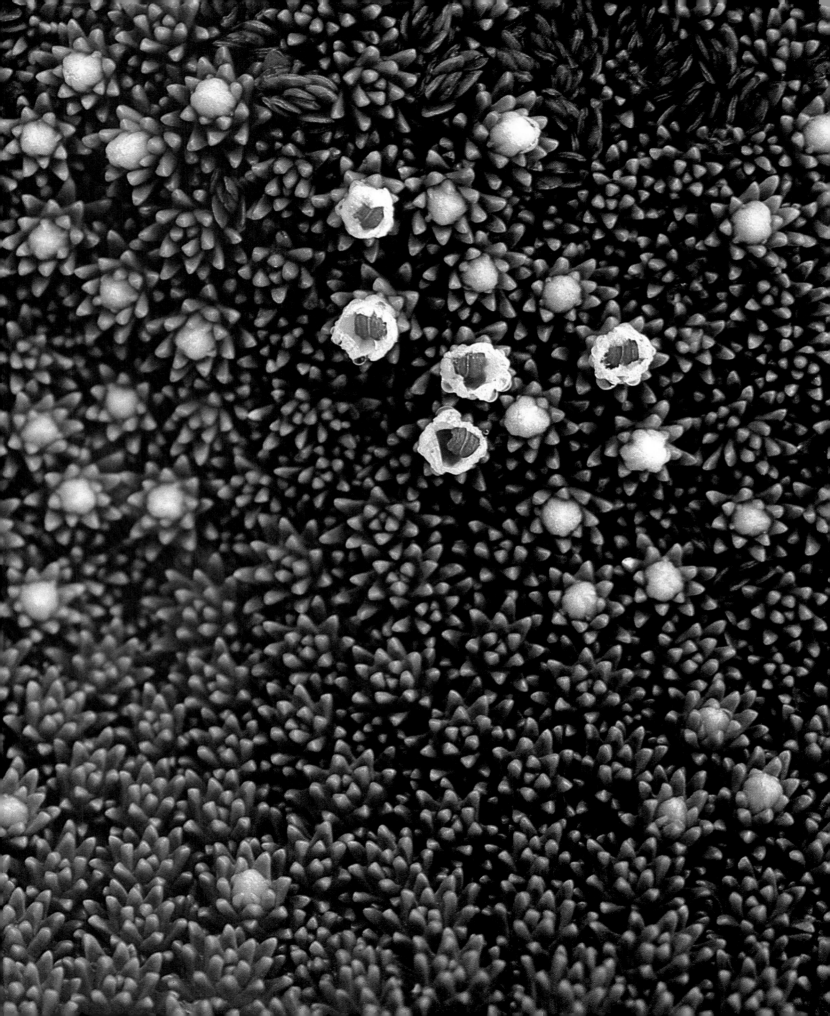

Permanent water and the protection of high cliffs in the gorges of the Arnhem escarpment give refuge to several of the hardier refugees of Australia's ancient pan-continental rain-forests. One of the commonest of these, Allosyncarpia ternata, is a descendant of the gum tree's forest ancestors. A small forest of them lines the walls of Jim Jim Gorge, in Kakadu National Park.

(RIGHT): Australia's eucalypts probably evolved from a rain-forest tree that looked rather like this. Although a member of the same family as the gum tree it belongs to the much older genus, Syzygium. Species of Syzygium are now widely scattered through the Old World tropics as well as Australia and the Pacific region, and they include one whose aromatic flower buds have

become known all over the world as a food spice. We call them cloves.

(OVERLEAF):Several eucalypts that live in Australia's arid areas have evolved a smooth, white bark that resists pests and reflects the heat of the grassfires that commonly sweep through their habitats. One of the most graceful of these is the Snappy Gum.

Syzygium alatoramulum, ATHERTON TABLELAND, QLD.

One of the more spectacular illustrations of this occurs on the heights of the Bellenden Ker Range, south of Cairns in Queensland, where Australia's only native Rhododendron grows in small isolated colonies. The family's ancestral home is in India where, in the foothills of the Himalayas, they form forests of great age. With the original family confined to temperate and mountain climates, the Queensland group could not have arrived by recent migration through the islands of Indonesia: it represents instead the last survivors of an ancient and widespread Gondwanan group. Rhododendrons would have evolved in India just as it was breaking away from Antarctica. They must have then dispersed through Antarctica to Australia and eventually become isolated in north-eastern Queensland when the rest of the continent dried out and became too hot for them.

The main threat to Australian plants and animals was aridity. Already toughened to some degree by poor Gondwanan soils, plants began to evolve new defences such as a thicker leaf cuticle and tighter breathing controls, to provide insulation and conserve moisture. This modified vegetation first appeared at the rainforest fringes forming a kind of closed but dry forest, known as Heidewald. Some scraps of it still survive in Australia. They are confined to the northern end of the Cape York Peninsula in the Jardine River basin. Other examples occur in New Guinea.

These new forests were the birthplace of that ubiquitous Australian, the gum tree. They now grow throughout Australia: from the monsoonal north to the alpine heath of Tasmania, and from the harshest central desert to the rain-drenched mountain forests of the Great Divide.

There are between 500 and 600 species of eucalypts, depending on the idiosyncrasies of the classifier. Their variety and restricted distribution guarantees that both they and their large sister group, the Leptospermum alliance, which includes tea trees and paperbarks, both evolved in the Australian region. Their family, the Myrtaceae, is clearly an old Gondwanan one.

Allosyncarpia ternata, JIM JIM GORGE, KAKADU, N.T.

201

TRUNK LINES

The gum tree and its relatives form Australia's second largest family group, the Myrtaceae. As an ancient Gondwanan family it adapted particularly well to Australia's poor soils and erratic climate, with eucalypts evolving from a rainforest fringe-dweller more than 30 million years ago.

Many are resistant to both drought and fire and they are well adapted to the poor soils of Australia's heaths and woodlands. Eucalypts dominate in all but rainforest and desert regions. Their adaption to a wide variety of habitats has produced a spectacular array of forms that ranges from stunted alpine species to the world's tallest flowering plants.

WESTERN SNAPPY GUM, *Eucalyptus leucophloia*, W.A.

BLUE GUMS, *E. deanei*, BLUE MTS., N.S.W.

SNOW GUM, *E. pauciflora*, FALLS CREEKS, VIC.

KARRI GUMS, *E. diversicolor*, NORTHCLIFFE, W.A.

SALMON GUM, *E. salmonophloia*, KELLERBERRIN, W.A.

GHOST GUM, *Eucalyptus papuana*, N.T.

BAXTER'S KUNZEA, *Kunzea baxteri*, MT. RAGGED

YELLOW MOUNTAIN BELL, *Darwinia collina*, STIRLING RA.

A CLAWFLOWER, *Calothamnus sp.*, STIRLING RA.

ALL IN THE FAMILY

One of the largest and most varied flower groups belongs to the Australian branch of the ancient Myrtaceae family.
While the plants themselves seem, in many cases, to have little in common, in their flowering structure the family resemblance is strong. All share three characteristic elements: a stout flower cup with detachable cap, and a profusion of long, showy stamens – elements flamboyantly defined in the tri-coloured blossoms of the Illyarrie Gum. The Illyarrie, and all the related forms shown here, are endemic to the south-western corner of the continent, for as is the case with many of Australia's old southern plant families, this is the centre of diversity for the Myrtaceae family also.

206

CORKY HONEYMYRTLE, *Melaleuca suberosa*, W.A.

YELLOW MORRISON, *Verticordia chrysantha*, PERTH.

ILLYARRIE GUM, *Eucalyptus erythrocorys*, W.A.

SWAMP BOTTLEBRUSH, *Beaufortia sparsa*, MARGARET R.

GIDGEE, *Acacia cambagei*, S.A.

Acacias, or wattles as they are commonly called, represent one of the most successful of the early flower groups to appear in Australia. While other members of the Acacias' family are widely dispersed throughout the world, particularly in the northern hemisphere, three quarters of the world's Acacias are Australian. They have adapted well to most habitats but their greatest success has been in arid areas where they often dominate, thriving in some cases where little else will grow.

WATTLE, *Acacia dunnii,* N.T.

PAPERBARK TREE, *Melaleuca sp.*, C. ARID, W.A.

It was the gum tree, however, which proved most versatile genetically, evolving a wide variety of trunks, leaves, flowers, fruits and growth habits. This diversity enabled them to cope with almost anything the fluctuating climate could throw at them. One of the group's more remarkable survivors is the Snowgum of the Australian alpine regions. Despite being alternately encased in winter ice and grilled by summer sun, some old Snowgums have survived hundreds of years, often with their hearts burnt out by the fires that occasionally sweep through the heavily wooded hills.

Two other spectacularly successful plant groups that diversified during this time were the Acacias and the gaudy Australian branch of the African Protea family. Both families have a basically Gondwanan distribution but include several northern hemisphere connections, suggesting a very early evolution for their ancestral forms.

There are more species of Acacia in Australia than those of any other genus, and they account for 600 of the 800 species that have so far been described in world collections. The Australian branch of the Protea family is much smaller but extraordinarily varied. It includes such groups as Banksias and Grevilleas, Hakeas, Dryandras, Bottlebrushes and many others. Their common ancestor appears to have evolved in an African rainforest about 100 million years ago and its descendants began to disperse into the Australian region between 10 million and 20 million years later. Banksias and Grevilleas proved to be the most successful colonisers, the versatile Banksia becoming one of the few groups in the world to adopt almost all plant forms, from full-sized tree to underground creeper.

PAPERBARKS, *Melaleuca preissiana*, ALBANY, W.A.

Although Australia's paperbark trees are linked to eucalypts within the Myrtaceae family, their growth habits are very different. Where the gum tree tends to be drought adapted and smooth limbed, paperbarks are characterised by the multilayered sheath of old bark that clings to their trunks. Like the eucalypts, most can regenerate easily from what are known as lignotubers, a concentration of specialised cells at the junction of roots and stem that often develops into a bulbous, knotty bole.

211

GREVILLEA, *Grevillea goodii*, TABLETOP RA., N.T.

ACORN BANKSIA *Banksia prionotes*, W.A.

RED POKER, *Hakea bucculenta*, PERTH, W.A.

GREVILLEA, *Grevillea miniata*, KIMBERLEYS, W.A.

SHOWY DRYANDRA, *Dryandra nobilis*, W.A.

Protea sp., SOUTH AFRICA.

ROSE CONEFLOWER *Isopogon formosus*, W.A.

SCARLET BANKSIA, *Banksia coccinea*, W.A.

WARATAH *Telopea truncata*, MT. ELIZA, TAS.

UNDERGROUND BANKSIA *(undescribed species)*, W.A.

MOUNTAIN ROCKET, *Bellendena montana*, WESTERN TIERS, TAS.

THE PROTEACEAE, A GONDWANAN GIFT

Of all the continent's flowers, the most flamboyant and characteristically Australian belong to a family of plants whose ancestors were clearly Gondwanan. They are known as the Proteaceae after their African relatives, the Proteas. The common ancestor of both branches of the family seems to have spread rapidly throughout Gondwana some 90 million years ago. Their success was due – in part at least – to their peculiar adaptation to bird and animal pollinators.

Most diverse among this highly successful group are the banksias. The genus has the rare distinction of including all three major forms – tree, shrub and prostrate – among its species. Some even grow almost entirely underground, showing only their leaves and flowers.

213

ROYAL HAKEA *Hakea victoria*, W.A.

Most spectacular of the hakeas is the Royal Hakea that inhabits a small region of heathland centred on the Fitzgerald River National Park in the far south of Western Australia (BELOW). As the big spiky leaves mature, over several years, their colour changes from pale green and yellow to crimson and purple, before they die and fall.

The Macadamia tree (BELOW RIGHT), one of the most primitive of the Protea's Australian relatives, also provides Australia's only cash crop from a native plant. Recent dispersal from its Australian birthplace has taken it to several south-east Asian islands.

MACADAMIA, *Macadamia integrifolia*, QLD.

NUMBAT, *Myrmecobius fasciatus*, DRYANDRA, W.A.

The whole Protea family is particularly adapted to pollination by birds and animals rather than by insects. The flowers are large or in clusters and they are borne on stout woody stems, strong enough to support animal pollinators. That these specialised plants thrived indicates that they brought their pollinators with them to Australia.

We have a good idea of what some of their mammal pollinators looked like because one of them still thrives in the flower-filled forests and heathlands of Australia's south-west. It is the Honey Possum. Despite its name it is not a member of the possum family but a lone survivor of a far more archaic group whose other members are now extinct. Equipped with a long, feather-edged tongue and a highly specialised digestive system, the Honey Possum is the only non-flying mammal in the world that still lives almost exclusively on a diet of nectar and pollen.

Two other evolutionary relics whose ancestral lines diverged during Australia's isolation are the termite-eating Numbat and the rabbit-eared Bilby or Dalgyte. Both were once widespread but now live near to extinction: the Numbat in the wandoo forests of south-western Australia and the Bilby in the arid regions of the central north and north-west of the continent. Like the Honey Possum, they too have no close modern relatives and their family connections have been blurred by time.

(RIGHT): Well satisfied after a feed of pollen and nectar, a young Honey-possum dozes on a dryandra flower in the far south-west of Western Australia. Like the Numbat and Bilby, the Honey-possum is the last survivor of its ancient family, which probably provided many such pollinators for the flower-filled forests that once covered much of Australia. The termite-eating Numbat (ABOVE LEFT) is now restricted to a small range in south-western Australia, and the Bilby (BELOW LEFT) to parts of northern Australia that are beyond the reach of the rabbit.

BILBY, *Macrotis lagotis*, ALICE SPRINGS, N.T.

HONEY POSSUM, *Tarsipes rostratus*, ON *Dryandra quercifolia*, W.A.

SIDE-NECKED TURTLE, *Pseudemydura umbrina*, W.A.

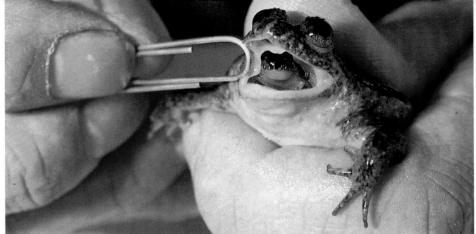

GASTRIC BROODING FROG, *Rheobatrachus silus*, QLD. PHOTO: MIKE TYLER, ADELAIDE UNIVERSITY.

Pausing for a moment on its mother's tongue this baby frog surveys the world for the first time as it completes one of the world's most remarkable birth processes. It was both hatched and reared within its mother's stomach sac. Two closely-related species of Queensland frogs use the same reproductive channel, swallowing their eggs after fertilisation, and then giving birth orally to a clutch of twenty or so fully formed baby frogs some six to seven weeks later.

The frogs are remnants of an ancient Gondwanan group, and while other branches of the family are diverse and widespread in Australia and South America, these sole survivors of their genus are confined to two small mountain habitats some 800 kilometres apart on Australia's central east coast. Neither have been seen since 1985 and are probably now on the verge of extinction.

(LEFT): The West Australian Side-necked Swamp Turtle also faces extinction. It is one of two relict species whose only other family members live in New Guinea and South America. They provide yet another instance of wide dispersal by a Gondwanan ancestor and a family break up caused by continental drift and changing environments.

A similar evolutionary 'aristocracy' exists within other groups of backboned animals. Notable among the birds are the emu, the cassowary, the lyre bird, the bower bird, and mound builders such as the mallee fowl. All have close family ties with living species on other continents that were once part of Gondwana.

Perhaps the most remarkable among this southern aristocracy are Australia's two aquatic frogs that were discovered recently in Queensland. These related species were found in two widely separated mountain communities along the Great Dividing Range. The first, *Rheobatrachus silus,* was collected in 1972 in the Conondale and Blackall Ranges north of Brisbane. The second, *Rheobatrachus vitellinus,* was discovered in 1984 in the mountain creeks of Eungella National Park, west of Mackay. Females of both species swallow their eggs soon after they lay them, and the young grow into froglets in the mother's stomach. They leave the way they enter—through her mouth. It is a reproductive pathway used by no other creature in the world.

Studies of *R. silus* revealed that the eggs, tadpoles and froglets all release a chemical that both inhibits the production of the mother's digestive juices (mainly hydrochloric acid) and switches off the powerful muscular contractions of the stomach wall that normally occur during digestion. This turns an otherwise normal frog stomach into a suitably inert and protective egg sac.

Strangely the other species, *R. vitellinus,* employs a slightly different chemistry. In this case the eggs and hatchlings are given a liberal coating of a special mucous to protect them from the hydrochloric acid in the digestive juices.

Birth is oral in both species. When the froglets are ready the mother surfaces, opens her mouth and they hop out, one by one. This birth process is normally spread over several days but if alarmed the mother may eject the entire brood in a form of projectile vomiting that lasts barely two or three seconds. Since no digestion can occur during the brooding phase, the mother cannot eat during this period. In fact, in later stages of the incubation process the stomach becomes so

Among Australia's most primitive trees is this rare tropical species (RIGHT) recently identified in the coastal rainforests of north-eastern Queensland. It is the only surviving member of an ancient and primitive family whose relationships are ill defined. Bearing heavy seeds the size of tennis balls, the species is almost immobile and clearly represents an old Gondwanan family that has otherwise become extinct. This and other discoveries have forced a re-evaluation of the old belief that most of Australia's tropical vegetation arrived as invaders from South-east Asia.

Idiospermum australiensis, DAINTREE, QLD.

distended by its load of froglets that it compresses and pushes aside the other internal organs. The largest recorded brood is 26 froglets.

Tragically, neither species of Gastric Brooding Frog has been seen since 1985. A severe drought and logging in parts of their habitats may have pushed both to the verge of extinction.

Environmental pressure is both a reaper and sower of species, and while the casualty list was long during Australia's voyage north, it was balanced by the even longer catalogue of new species that flourished. Prominent among them were two of the world's major mangrove types, the family Rhizophoraceae and the genus Avicennia, which seem to have originated at very different times along the northern or western margins of the Australian plate. They developed from separate species of land plants as each learnt the trick of purifying saltwater.

The Rhizophora family evolved first—probably about 30 million years ago—from the same general group of plants that the gum tree was evolving from at that time. Rhizophora seeds are buoyant and designed to survive long sea voyages. Distributed by the dominantly east-west equatorial weather patterns, they spread swiftly throughout Asia. Africa had not yet closed with Europe and the gap allowed Rhizophora to slip through the narrowing Mediterranean corridor to the Atlantic and eventually to the Americas. It is the nearest thing to world conquest by any Australian plant family.

SPIDER MANGROVE, *Rhizophora stylosa*, QLD.

The worldwide mangrove family known as Rhizophoraceae seems to have evolved in the Australian region around the same time as the gum tree, and from a relative of the same family. The Rhizophoras, a major branch of the family, are characterised by looping prop-roots (BELOW).

Clear evidence of the existence of Gondwana, and of its dismemberment, is provided by a remarkable group of spiders. Known as Retarius, or Net-casting Spiders, these are found in many coastal regions of Australia. Their closest relatives however, live in Central and South America. Their most notable feature is a pair of huge hunting eyes,

wholly adapted to night vision. But even more remarkable is the carefully crafted 'hand' weapon with which they catch their nightly dinner. Using a back foot, these spiders 'knit' their multi-stranded web-silk into a lacy elastic tape, and then form this tape into a small, rectangular catching net. Suspended over a carefully selected ambush area,

and holding the net with its four front feet, the spider is then ready to dab its net on to any suitable prey that passes beneath it. (RIGHT): A similar weapon is used differently by a very rare relative of the Tasmanian Cave Spider. This species, the Carrai Cave Spider, does not attack with the net but flicks victims into it with its two long front legs.

NET-CASTING SPIDER, *Dinopis subrufa*, N.S.W.

Other notable Australians that appeared during this period of major environmental change were the casuarinas or she-oaks, and the curious family known as grass trees. The main group, Xanthorrhoea, are the highly modified progeny of a primitive Gondwanan member of the lily family. The casuarina's connections, on the other hand, are not yet well understood but there is evidence to suggest that they may be very early off-shoots of one of the oldest flowering plants of all, one that lay somewhere between the primitive catkin bearers, such as willows, and the earliest of the true flower bearers.

NET-CASTING SPIDER, *Dinopis subrufa*, N.S.W

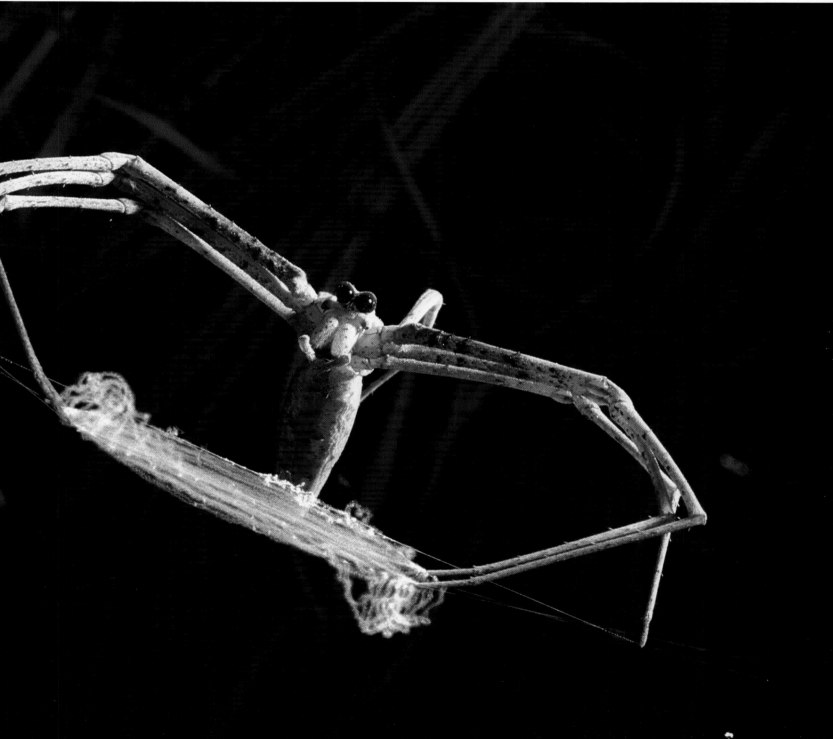

DESERT OAK, *Casuarina decaisneana*, N.T.

(LEFT): The bulbous Boab trees of the Kimberleys are closely related to Africa's giant Baobabs. The similarity between these two drought adapted species suggests that their common ancestor differed little and dispersed widely through the arid regions of northern Gondwana before it fragmented.

(ABOVE RIGHT): The origins of Australia's casuarinas, or she-oaks, remains obscure but they appear to be derived from some of the very earliest flowering plants, the catkin bearers such as willows and liquidambars. Australian species have adapted to a wide range of habitats, the Desert Oak for example providing the only shade there is in many parts of inland Australia.

(RIGHT): The State emblem of Western Australia, the red and green version of Mangle's Kangaroo Paw, belongs to another old southern family, in this case centred on Western Australia although it can claim relatives in Africa as well as Central and South America.

(BELOW RIGHT): Only one member of Australia's ancient southern pines has managed to meet the challenge of current aridity. This is the White Cypress Pine of inland and northern Australia, one of 13 species in its genus and the hardiest and most widespread.

KANGAROO PAW, *Anigozanthos manglesii*, W.A.

WHITE CYPRESS PINE, *Callitris columellaris*, N.T.

BOAB, *Adansonia gregorii*, W.A.

225

GREAT SOUTHERN ARK

THE LILY CONNECTION

Australia's unique grass trees are a bizarre offshoot of the ancient lily family. They grow in scattered pockets right across southern Australia. It is a pattern that suggests they were once part of a widespread forest family that became carved into disjunct populations by environmental change.

The family is now represented in Australia by some 70 species that fall into nine groups, or genera of which only two are generally recognised as grass trees. Of these, the Xanthorrhoeas are best known because of their curious leaf crown and spectacular flower spikes. Like several of Australia's hardy old residents they have evolved species that now occupy a remarkably wide variety of habitats, ranging from semi desert to cold, wet coastlines and frosty mountaintops.

Xanthorrhoea sp., BUNYA MTS, QLD.

Xanthorrhoea sp., CAPE OTWAY, VIC.

Xanthorrhoea sp., GOSSE BLUFF, N.T.

This forest of grasstree flower spikes (BELOW) had been stimulated into growth by a recent bushfire. Such 'bounce-back' mechanisms as this are a hallmark of most of the major families of plants and animals that evolved in Australia during its long solo voyage to the tropics.

LAVA TUBE, UNDARA CRATER, QLD.

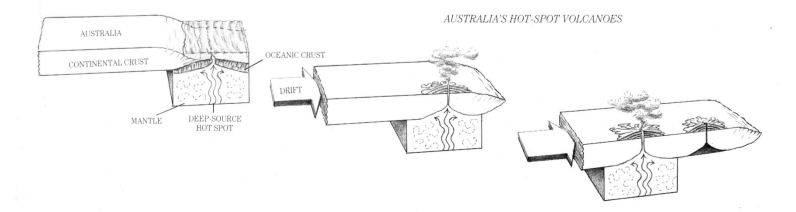

AUSTRALIA'S HOT-SPOT VOLCANOES

AUSTRALIA

CONTINENTAL CRUST

OCEANIC CRUST

DRIFT

MANTLE

DEEP-SOURCE HOT SPOT

VOLCANIC MILESTONES

Some of eastern Australia's most memorable scenery is traceable to several stationary hot-spots deep in the Earth's mantle. They underlie Australia's continental raft and have periodically burnt their way right through it to produce a series of volcanoes that now mark Australia's steady northward drift.

Typical of these volcanoes are the Glasshouse Mountains of south-eastern Queensland. They consist of the plugs of lava that solidified in the throat of the vents after the final eruption (BELOW). Four million years later, when the Australian raft had drifted 150 kilometres further north, the same deep-seated hot spot again ate its way through the continental crust producing a flood of lava. The vent is now represented by the big volcanic plug known as Mount Warning (ABOVE RIGHT) on the northern border of New South Wales.

Hotspot volcanoes are not things of the past in Australia. An eruption that occurred only about 95,000 years ago near Mount Surprise in north-eastern Queensland has left a cluster of small craters and one of the longest sets of lava tubes in the world (LEFT).

Such tubes occur where smooth, fluid lavas form long rivulets. The surface material cools and hardens, insulating the internal lava which continues to flow. When the eruption ceases the tubes empty and remain hollow.

MOUNT WARNING, N.S.W.

With Australia attached to the south-eastern edge of the Indian oceanic plate, the two continents continued to drift northward at about 50 millimetres a year. India began to plough into southern Tibet when Australia was still more than 500 kilometres away from the chain of Indonesian islands known as the Sunda arc. The rest of Australia's journey to the inevitable collision with the Indonesian islands is accurately paced off by the volcanoes that began to puncture its eastern side. They were punched through the continental crust by deep-seated mantle 'hot spots' as the Australian raft began to pass over the simmering remnants of the crustal rift that had so recently opened the Coral Sea. It was like passing a sheet of iron over a blowtorch: every now and then the heat built up below the continental slab to such a degree that a new hole burnt through, and another volcano was born.

Australia's hot-spot 'holes' run in three parallel lines and they range in age from 32 million years old near Nebo, Queensland, to five million years old near Macedon in southern Victoria. These mantle hot spots were not the only cause of recent Australian volcanoes, but they were a major source of lava, producing at least 16 large shield volcanoes. 'Shield' refers to the low-profile, lava dome produced by the smooth-flowing magma that characterises such deep-source eruptions. It is like an Aboriginal tribal shield placed face-up on the ground. Linked together on a map, these punctures precisely chart Australia's northward passage during its inexorable collision with Indonesia. As the big Australian raft began to shoulder its way into the necklace of southern islands, a trickle of plants and animals began to cross the narrowing water gaps in both directions. Australia's genetic isolation was coming gradually to an end and its role as an ark was drawing to a close.

MT. COONOWRIN, GLASSHOUSE MTS., QLD.

20 – 5 million years

WITH SIX MINUTES *left on the time scale, Australia comes to the end of its long solo voyage. The northern edge of the continental raft ploughs into the chain of volcanic islands that line the south-western rim of the Pacific Ocean plate, sweeps them aside and begins to override the plate itself. Crumpled and deformed, this northern edge of the continent rises slowly out of the sea to form the New Guinea highlands. The land behind it subsides into the Arafura Sea. The exchange of land life with South-East Asia is the first such interchange since Australia left Antarctica. With about four and a half minutes to go, polar temperatures take another plunge. Grasslands spread through the world's temperate regions and large grazing animals proliferate. Like their northern counterparts, Australia's marsupial grazers are generally large and highly mobile. Among these are the ancestors of modern kangaroos and wallabies, which can move about freely on their hind legs. Meanwhile in Africa a very different kind of two-legged mammal appears. Forced out of the shrinking forests, one of Africa's primates faces the hazards of the open plain by standing upright.*

These volcanoes were still smoking as Australia moved towards an inevitable impact with the Indonesian islands. By the time the mantle hot-spot that caused them reappeared, some four million years later, the continental raft was 150 kilometres further north, and the impact had begun.

GLASSHOUSE MOUNTAINS, QLD.

GATHERING STORM

THORNY DEVIL, *Moloch horridus*, N.T.

GATHERING STORM

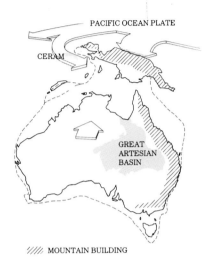

PACIFIC OCEAN PLATE

CERAM

GREAT
ARTESIAN
BASIN

///// MOUNTAIN BUILDING

(LEFT): Wedged in storm debris, tiny eggs such as this one dangling from the foot of a baby stick insect, would occasionally have rafted across water gaps to Australia, long before the main collision between Australia and the southernmost Indonesian islands began.

(ABOVE): Among the earliest land vertebrates to make the crossing were ancestors of this small, highly specialised desert lizard known as the Thorny Devil.

(OVERLEAF): Well adapted to the rigours of desert life, the Thorny Devil makes good use of every scrap of available water. It need only step into a mud puddle and a dark, glistening tide of moisture gradually envelops its entire body, eventually seeping directly into its mouth. In fact water need only touch its skin and its body 'drinks' through pores in the tiny canals that crease its thorny hide.

Behind the daunting 'thorns' that protect the Australian desert lizard *Moloch horridus*, there lurks a shy and gentle hermit. It needs only a clump of spinifex for shelter and a nest or two of small black ants—its exclusive diet—to find contentment. Even a regular supply of rainwater is unnecessary because the Moloch has almost forgotten how to drink in the normal way. An occasional stroll through the morning dew is now enough to keep it well watered.

By this spectacular adaptation to aridity, the Moloch sidestepped all the major problems of desert survival. The secret lies in the micro-structure of its thorny hide. Its keratin-coated skin is creased by a network of very small open canals. By a process known as capillary action, any water that touches a Moloch's skin will spread along these channels until all are filled. In this way the Moloch may drink merely by standing in a mud puddle. Within a minute or so, water seeps up its legs and over its entire body. As the tide of moisture creeps over its face, the Moloch gently opens and closes its mouth, and the water finds its own way in. It is also able to absorb some water directly through its skin, a talent more common in amphibians such as frogs and salamanders.

For a lizard whose family ties connect it with monsoonal Asia, such extreme modification indicates a long residence in Australia's arid heartland. The Moloch's ancestors must have arrived early, spread across the continent, and then kept pace with the drying out by a gradual process of adaptation through natural selection. Its Asian ancestors were probably rafted here some 15 million to 20 million years ago on storm debris, as Australia approached the chain of islands that fringed the westward moving Pacific Ocean plate. As one of the earliest of the northern invaders, the Moloch is a memento of the end of the voyage of the great southern ark.

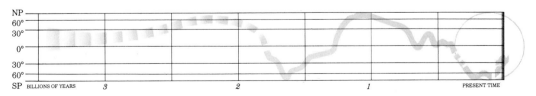

NP
60°
30°
0°
30°
60°
SP BILLIONS OF YEARS   3   2   1   PRESENT TIME

FOSSILISED BIRD, RIVERSLEIGH, QLD.

The Riversleigh fossil deposits that lie beside the Gregory River in north-west Queensland have more than doubled the catalogue of Australia's prehistoric animal species. The fossils range in age from about 15 million years to less than four million and occur in well-preserved deposits of bone and teeth in several outcrops of crumbling limestone (ABOVE). Many of those exposed to weathering have been etched into relief by the carbonic acid in rain and, like these leg bones of a giant bird (LEFT), now stand out clearly.

(BELOW): In its outward similarity to the giant flightless birds that once roamed Australia's plains, the emu appears to illustrate convergent evolution rather than a close genetic relationship. It is not known whether, among the giants of old, the males assumed the role of incubator as is the case with emus. This male is in the process of carefully turning the eggs.

More concrete evidence of early invasions by animals from the north has been discovered beside the Gregory River in north-western Queensland. Here, on Riversleigh station, the remains of thousands of animals of many kinds have been found, embedded in limestones that range in age from 24 million years down to about 10,000 years. Many of these fossils have Eurasian family links, such as bats, rodents, snakes, lizards, crocodiles, and several frogs. However, bat fossils are by far the most numerous and of the 25 or more species so far recognised, several show close ties with contemporaries that lived as far afield as north Africa and France.

One of the surprising features of Riversleigh is the richness of the old southern fauna. Inhabitants of what was then a region of tropical wetlands included the leopard-like *Wakaleo*, a small Tasmanian 'tiger', a very large toothed platypus, three species of lungfish, a miniature koala, giant marsupial browsers and, in the trees, possums of many kinds. Among them were many animals new to science, whose recent discovery has more than doubled Australia's official catalogue of prehistoric mammals. Outstanding among these new forms are a group of small kangaroos whose jaws and teeth are unmistakably those of meat-eaters. An entirely new order of mammals has also been recognised. This group, affectionately nicknamed 'Thingodonta', represents

EMU, *Dromaius novaehollandiae*, S.A. (PHOTO: JEAN-PAUL FERRERO, AUSCAPE)

DROMORNITHID FOSSIL, RIVERSLEIGH, QLD

one of the most distinctive fossil finds in Australia this century, although little as yet is known of its general structure or habits.

Riversleigh's largest mammal-remains belong to two groups of four-legged marsupial browsers, one of which reached the size of a rhinoceros. The bear-sized *Neohelos* and the even larger *Bemartherium*, represent the beginnings of a succession of marsupial giants that dominated the Australian landscape until about 30,000 years ago.

Other giants represented at Riversleigh belonged to a family of huge flightless birds that roamed the woodlands and the edges of the rainforest. This group, the dromornithids or Mihirung birds, included one as yet unnamed species that was amongst the world's heaviest birds. It is also a contender for the title of the world's largest, growing three metres or more in height and weighing about half a tonne.

Though the dromornithids bear a superficial resemblance to both the emu and the cassowary, they are evidently not closely related to either.

The extraordinary melange of fossilised life that became entombed in Riversleigh's limestones offers not only a unique peephole into the past but a view of a crucial event in modern evolution. It represents the

This is the leg bone and the gizzard stones of a giant bird known as a dromornithid. Worn smooth as though polished in a gemstone tumbler, these stones provided the dromornithids with a substitute for teeth by helping to grind down their coarse foods within the stomach sac. Many modern seed eaters, notably emus, still employ the same device.

Palorchestes azeal

A FLOURISH OF GIANTS

DRAWINGS: PETER SCHOUTEN,
'Prehistoric Animals of Australia'

The onset of the present ice age produced a flourish of giant land mammals all over the world. Australia was no exception. By five million years ago all its current marsupial groups were well established. The large ancestral family known as diprotodontids were still dominant, however, grazing and browsing throughout Australia's savanna, woodland and open forests.

Gigantism had appeared early in this group with the bear-sized *Neohelos* and the even larger *Palorchestes,* which looked something like the tapir of South America. Both failed to survive the aridity that set in at this time but among the diprotodontids that replaced them was the largest marsupial that ever lived, *Diprotodon optatum*

Three metres long and two metres high at the shoulder, this species appears to have thrived in woodland areas until 20,000 years ago or even less. Slightly smaller relatives occupied the forest regions.

A more bizarre member of the family was the large, pouch-cheeked *Euryzygoma,* which sported bony projections on each side of its huge head, just behind the eye sockets. These projections may have been solely there to support large cheek pouches, though they might have doubled as weapons or sexual ornament. Modern American gophers have similar supports for their cheek pouches. *Euryzygoma* first appeared about four million years ago and seems to have disappeared again about three million years later.

Euryzygoma dunense

Neohelos sp.

GREEN PYTHON, *Chondropython viridis*, QLD.

coming together of two streams of life that had been separated for almost 200 million years—even since Pangaea split into its northern and southern halves.

As the Australian continental raft continued to push through the chain of islands that fringed the edge of the Pacific Ocean plate, its northern rim began to override the plate itself. Amid a crescendo of crustal shudders and volcanic eruptions this rim began to buckle and lift, heaving upwards to form the sharp spine of the New Guinea Highlands. Counterbalancing this uplift, the land immediately south of

Pythons, lizards and bats were among the most successful of the early Asian invaders. One rare and handsome descendant is the Green Python (ABOVE). It is confined to the rainforests of Cape York Peninsula, and while its adult colouring provides excellent camouflage, juveniles

240

LITTLE RED FLYING-FOX, *Pteropus scapulatus*, N.T.

FRILL-NECKED LIZARD, *Chlamydosaurus kingii*, N.T.

may range from bright yellow to rusty red.

Most numerous of the northern invaders were the bats. Adapting readily to the nectar and pollen of Australian trees, especially eucalypts, fruit bats such as the Little Red Flying-Fox (*ABOVE RIGHT*) thrived here, forming huge colonies.

One of Australia's most spectacular lizards, the Frill-Necked Dragon (*RIGHT*), seems to owe its existence to an ancestor that entered Australia from the north soon after the continent began its collision with the Indonesian island chain.

241

CHILDRENS PYTHON, *Liasis childreni*, LITTLE BENT WING BAT, *Miniopterus australis*, QLD.

CHILDRENS PYTHON AND LITTLE BENT-WING BAT.

GREEN TREE FROG, *Litoria caerulea*, QLD.

AMBUSH AT BAT CLEFT

With no herd animals and no large carnivores in modern Australia the large-scale predator-prey interaction that is a feature of the plains of Africa has not developed here. The nearest thing to it occurs each summer at Mount Etna on Queensland's central coast. About 100,000 female Bent-wing Bats congregate at a single maternity cave to give birth and nurture their young. It is the signal for a nightly gathering of predators – pythons and frogs – that has no equal in this country. Up to 30 pythons distribute themselves around the walls of the cave's slot-like entrance and extend their necks into the torrent of bats that stream from its recesses. When a bat bumps into a snake it clings momentarily to re-orient itself (FAR LEFT) – just long enough for the python to throw a fatal coil over it.

Bats that settle, however briefly, around the lip of the cave run the risk of being snapped up by one of a small army of frogs that gather to pick up such 'crumbs' (LEFT).

the highlands warped downwards. Large areas submerged, leaving New Guinea separated from the rest of the continent by the Arafura Sea, Torres Strait and the Gulf of Carpentaria.

As the Australian raft punched through the chain of islands they swirled like floating leaves around New Guinea's western end and closed in behind its anvil head as though caught in a gigantic eddy. The 'swirl' is still visible in the shape of an 'S' whose upper loop of islands enfolds New Guinea's western end. The large northern island, Ceram, which entered the eddy most recently, is now completely reversed. Having 'swirled' past New Guinea's western end, it has rotated through 180 degrees: its present southern shores originally faced north. Timor and the Aru Islands, on the other hand, were not influenced by this eddy. They represent the north-western edge of the Australian raft.

Australia's collision with the western edge of the Pacific Ocean plate and the chain of Indonesian islands that fringed it have left three kinds of scars along the northern edge of the continental raft. It raised the New Guinea highlands, submerged the vast forest-covered plains that lay to their south, and edged the Australian raft into the vast belt of volcanic unrest known as the Pacific Ring of Fire. These two dead volcanic vents form part of the Murray Island group at the eastern end of Torres Strait. They were active less than a million years ago.

A GARLAND OF ISLANDS

Australia begins to impact the chains of volcanic islands that define the edge of the westward-moving Pacific Ocean plate.

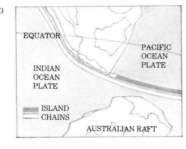

20 MILL. YRS. AGO

EQUATOR

INDIAN OCEAN PLATE

PACIFIC OCEAN PLATE

ISLAND CHAINS

AUSTRALIAN RAFT

10 MILL. YRS. AGO

The leading edge of the Australian continental raft pushes into the island chains, and where it meets the Pacific plate, it crumples to form the highlands of New Guinea.

PRESENT

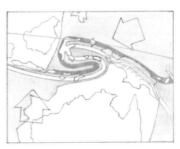

Still driving westward, the Pacific plate develops an 'eddy' known as a drag fold at the western end of New Guinea, looping the island chains around it like a twin-strand necklace.

ADAPTED FROM J. J. VEEVERS, 1984

Without this chain of island stepping-stones, few species could have crossed the water barrier between Australia and South-east Asia. The loop, which began to develop in the chain about 10 million years ago, served to multiply the number of migration routes. Australia's approach to the islands had been so slow that the two environments were not so very different at their junction when they met and many animal species would have found little difficulty adapting.

One surprising feature of the animal exchange was that physical mobility was not the governing factor in a species' dispersal: though

YELLOW-BREASTED SUNBIRD, *Nectarina jugularis* QLD.

many ground dwellers made the crossing, relatively few non-migratory birds did. The dispersals were clearly accidental and because birds enjoyed a mastery of their environment they were much less vulnerable to such accidents.

There were, however, a few spectacular exceptions, one being the ancestor of the modern crow. The comparative analysis of bird DNA has exposed new patterns of evolutionary relationship. It indicates that all the world's ravens and crows, as well as several other major groups, evolved from an Australian ancestor which 'escaped' to Asia some 20 million to 30 million years ago. This was well before Australia collided with the island chain. Meanwhile, the same ancestral line in Australia spawned a wide variety of descendants, including magpies, currawongs, butcher birds and the New Guinea birds of paradise. Ironically Australia's modern ravens and crows are not among them. They evolved instead from descendants of the original 'escapees' which later found their way back to Australia—a return of the prodigal DNA.

Most of the similarities that exist between Australian birds and those of Africa, Europe and America, are not the result of common ancestry, as was first thought. They are a striking illustration of what is known as convergent evolution, in which animals with similar ecological roles evolve similar forms, despite genetic and geographic differences.

It was long believed that most Australian birds were derived from Old World families. DNA analysis is now revealing this to be largely untrue. For example, not only are Australian wrens (BELOW) unrelated to their Old World look-alikes, they come from an old Australian stock that also gave rise to many non-Australian families, birds including the world's magpies, ravens, jays, orioles, and even the common crow. They appear to have evolved elsewhere from some early Australian 'escapees'.

In Australia, meanwhile, another branch of the same stock was giving rise to such notable natives as the Great Bowerbird (RIGHT), shown removing an offending blossom from the grey-and-white decor of the display area in front of its bower.

By contrast, the Yellow-breasted Sunbird (ABOVE LEFT) resembles some Australian honeyeaters, but it is unrelated and a recent immigrant. Its family appears to have originated either in Asia or in Africa.

VARIEGATED FAIRY WREN, *Malurus assimilis*, S.A.

FLOATING CONGREGATION

Australia once had huge shallow lakes that provided permanent habitats for multitudes of birds, especially waterbirds. The lakes have vanished now but sufficient standing water remains to support a large population of itinerants. The largest of these flocks occur in the seasonal wetlands of the far north, where waterbirds such as whistle ducks and pied geese congregate in huge colonies (ABOVE RIGHT). In arid regions however, it is the seedeaters that form the largest flocks. So large are the flocks of Corellas that roost beside some inland billabongs that the trees appear to be covered in white blossom (ABOVE).

Unlike most other continents Australia has few major bird migrations that are truly seasonal and only along the coastlines do the regular travellers of the bird world appear in any numbers. These terns and sandlings (RIGHT) winter in the northern hemisphere.

MAGPIE GEESE, *Anseranas semipalmata*, N.T.

Though no other Australian animal appears to have achieved the worldwide spread of the ancestral crow, there are a few families, originally Australian, which now have a wide tropical distribution. The world's sea snakes for example, seem to fit this category and probably evolved from an aquatic form somewhere in northern Australia, a little more than 10 million years ago. Since they now occur as far west as the Caribbean Sea, the latest timing of this dispersal is limited by

TERNS AND SANDLINGS, 80-MILE BEACH, W.A.

LITTLE CORELLAS, *Cacatua sanguinea*, BIRDSVILLE, QLD.

Africa's collision with southern Europe, which closed off a pre-Mediterranean sea corridor around that time.

Most land mammals, unlike reptiles, make poor sea travellers and the island-hopping passage between Asia and Australia remained closed to them for a further five million years. The key to unlock that passage lay thousands of kilometres away in the polar icecaps. These grew rapidly during a major fall in temperature that began a little more than five million years ago. As the icecaps expanded they absorbed such huge quantities of the world's water that sea levels fell, exposing great

When sea levels were lowered by the growth of polar ice caps, it reduced the water barriers between Asia and Australia for plants and animals. The drop in polar temperatures that occurred about 14 million years ago would have considerably eased the exchange of species that was then beginning. One of the more attractive Asian invaders to arrive during the next few million years was the Barringtonia, or Freshwater Mangrove (LEFT). If it were not for the fact that the Barringtonia flowers mainly at night its peculiar significance as a colonist would be more noticeable. It is not by accident that the feathery blossoms of the Barringtonia bear a superficial resemblance to many eucalyptus flowers: they are in fact distantly related. The origins of both families can be traced to a common ancestor.

Old sea levels often leave their signature in the form of rock sculptures such as this limestone 'mushroom' (RIGHT) at Cape Keraudren in north-western Australia. Its upper surface represents part of a rock platform cut by the storm waves of 4,000 to 5,000 years ago.

LIMESTONE, CAPE KERAUDREN, W.A.

FRESHWATER MANGROVE, *Barringtonia acutangular*, N.T.

FOREST VINE, KIMBERLEYS, W.A.

RAINFOREST, FRASER ISLAND, QLD.

areas of land. The water barriers that had prevented all Asian mammals except bats from reaching Australia, shrank dramatically. Within a million years the first of the new northern forms, the placentals, had arrived. Predictably, they were the rodents.

Meanwhile the last of central Australia's rainforests had given way to savanna, grassland and desert. The tropically adapted immigrants, both plant and animal, became increasingly restricted to Australia's humid coastlines.

CLOUDFOREST, MOUNT SORROW, QLD.

GONDWANAN SURVIVORS

The growth of aridity in central Australia tended to restrict many rainforest plants, both Gondwanan and Asian, to Australia's humid coastal fringes. Though most Asian species were not well fitted to withstand either the cold or the poor soils that typified most high altitude habitats along the Great Dividing Range, many Gondwanan species adapted well to them. (ABOVE). But even at sea level there are many Gondwanan survivors.

Some rainforest species do survive right out in the open, however, where almost nothing else grows, and woody rainforest vines (TOP LEFT) still cling to a few stunted trees on baking Kimberley hillsides that are otherwise almost barren.

GATHERING STORM

THE GILDED CARNIVORES

Australia seems to be the home of two ancient families of carnivorous plants.

The Albany Pitcher Plant which probably originated on a cold rainforest floor, developed leaves that retained water in a jug-shaped hollow. They could then supplement their root intake by absorbing nitrogen and other nutrients produced by the bacterial decay of insects that drowned in these leaves. The Sundews, on the other hand, evolved leaf glands that exuded a sticky substance capable of immobilising insects that walked on them. Their modified leaves then absorbed the nitrogen in the residue of the insect's decay.

One insect has turned the tables on the Sundews. By controlled movements and powerful leg muscles a well-camouflaged Mirid Bug patrols the sticky leaves of its host plant with immunity (RIGHT). Where it finds an insect trapped by the Sundew the Mirid Bug skewers it with a needle-like proboscis and sucks the victim dry.

ALBANY PITCHER PLANT, *Cephalotus follicularis*, W.A.

SUNDEW, *Drosera indica*, HAMERSLEY RA., W.A.

MIRID BUG ON *Drosera macrantha*, W.A.

One Asian swamp plant has successfully colonised parts of Cape York Peninsula with the aid of the same carnivorous strategy as that used by the Albany Pitcher Plant, although they are entirely unrelated. Though its pitchers are similar, the Asian form (BELOW) has a very different growth pattern to cope with its warm, swampy habitat.

PITCHER PLANT, *Nepenthes mirabilis*, QLD

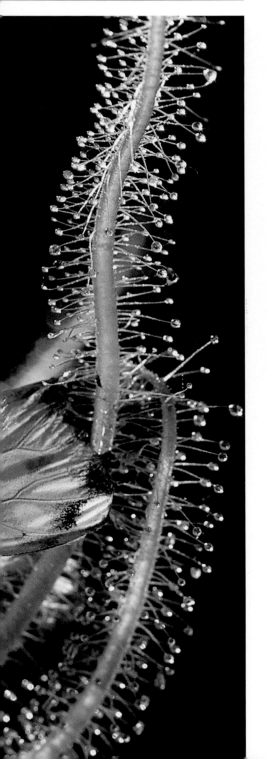

A SLY CONGRESS

Without doubt the most remarkable relationship between plants and animals occurs between an ancient group of southern ground orchids and a few species of wasp. In the baldest anthropomorphic terms, the orchids play the harlot by disguising themselves as female wasps and virtually raping the male wasps they attract.

The most spectacular of this group is the Hammer Orchid (BELOW) of south-western Australia. It baits the trap by growing a fair imitation of a female wasp on the end of a long, hinged armature. It also releases a sex scent, or pheromone, similar to that used by female wasps of the Thynnid family, to guide their suitors to them. The females of this group are wingless and normally wait at the top of grass stems for wandering males to abduct them for the mating process. However, a male that tries to carry off an orchid decoy in the traditional manner finds itself flipped on to its head and branded with the orchid's pollen sac.

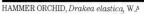
HAMMER ORCHID, *Drakea elastica*, W.A

TRIANGLE WITH A TWIST

Perhaps the most remarkable invader from the north is an aerial relative of the common Gardenia. Living high among the branches of paperbark and mangrove trees this spiny, ungraceful epiphyte has come to an arrangement with two species of small brown ants. The ants burrow out a home in its potato-like base and in return, protect the plant from other animals.

Such a partnership is by no means rare, but occasionally a third player enters the arena. A small blue butterfly is known to seek out these plants and lay its eggs nearby. The ants then carry the eggs inside their spiny fortress and store them until they hatch.

The ants allow the larvae to feed on the walls of their home and the larvae in turn exude a substance attractive to the ants. When they mature, the larvae eat their way to the outer skin of the plant where they pupate, and then fly away as butterflies.

The Anthouse Plant and its ants seem totally interdependent: neither can exist without the other.

ANTHOUSE PLANT, *Myrmecodia beccarii*, QLD.

EGGS AND PUPAE, *Iridomyrmex cordatus*.

(LEFT): One ancient group of plants evolved few of the usual defences, but thrived nevertheless with the aid of its sticky seeds and an obliging bird that fed on them. This group, a family of mistletoes, learned to live as parasites on other trees and depend on the Mistletoe Bird for seed dispersal.

MISTLETOE, *Ameyema sanguineum*, N.T.

BARKING SPIDER, *Selenocosmia stirlingi*, N.T.

GOULD'S GOANNA, *Varanus gouldii*, N.T.

THE DESERT UNDERWORLD

The onset of the current ice age forced many animals to look to their defences against the increasing environmental stress, and aridity in particular. For most small animals the best refuge lay underground. As the forests withered and the natural bolt-holes offered by tree stumps and fallen logs became scarce they learned to dig deeper. Apart from the occasional kangaroo or emu, and a few reptiles, Australia's arid regions now seem deserted by day. The underworld comes alive as the sun sets.

WATER-HOLDING FROG, *Limnodynastes spenceri*, N.T.

DESERT SCORPION, *Urodachus sp.*, W.A.

HONEY-POT ANTS, *Camponotus inflatus*, N.T.

THE SPECIALISTS

Two specialists of the desert under-world are an ant and a blind mole.

The Honey-pot Ant is so named because during the desert's periods of abundance workers force-feed nectar to certain members of the colony, turning them into living larders. These 'repletes' as they are called, form the colony's insurance against leaner times. (ABOVE). When a worker needs a feed it grooms one of the repletes which then responds by disgorging some of its honey into the worker's mouth

Remarkably, however, the major predator of the under-ground world is not only a marsupial, but one of the most bizarre. In adapting to a life spent 'swimming' through dune sand, its ancestors gradually lost their external eyes and most of their ears. They developed large, horny snouts, massive front claws and backward-facing pouches to protect their young. Alerted, perhaps, by vibrations trans-mitted through the sand, this mole has successfully attacked the gecko from below (TOP RIGHT).

MARSUPIAL MOLE, *Notoryctes typhlops*, N.T. (PHOTO: MIKE GILLAM, AUSCAPE)

There is one animal, however, that appears to have thrived on the expansion of Australia's sand dune systems. It is the elusive Marsupial Mole of central and north-western Australia. Among the assets it evolved for 'swimming' through dune sand were a sleek, mobile body without eyes or ears, well-developed front legs armed with massive claws for burrowing, and a backward-facing pouch to protect its teats and suckling young. It was once thought that its extreme specialisation was evidence for the long existence of a desert dune system, but an ancestral mole has turned up among the 15-million-year-old rainforest fossils at Riversleigh in Queensland. Like the Platypus, it is highly specialised and the sole survivor of an ancient mammal order.

While the well-watered eastern side of the continent remained the preferred highway south for most latecomers, few environments are truly invulnerable to a fast evolving modern animal group such as the rodents. As one of the world's most successful mammals, they penetrated almost every available habitat within the next few million years.

There would be only one invasion more dramatic. Its seeds were being sown even then among a group of primates, half a world away, on the drying plains of East Africa.

HONEY-POT ANT REPLETE AND WORKER, N.T.

5 million years – present

THE LAST 90 SECONDS *of the time scale begin amid climates much like those we live in today. Antarctica's ice sheet extends well out to sea for the first time and drought stalks the world's mid-latitudes.*

Because large bodies retain heat better than small ones, giant mammals thrive all over the world. In Australia, however, a lack of water limits this development, especially among the less mobile species. But large birds survive well. The giant flightless forms still roam Australia's plains, while huge flocks of pelicans, flamingoes, swans, ducks and other large water birds congregate about the shrinking remnants of its ancient inland lakes.

With 30 seconds to go, world temperatures take the final plunge and the ice age begins. Drought becomes commonplace, deserts expand and fire-resistant vegetation spreads. And in Africa an even more insidious threat appears. A new kind of hominid with a 'top-heavy' brain has just evolved. Wandering groups of them begin to leave their arid homeland, searching for greener pastures. They spread into Europe and Asia, just as the glacial cycles begin. Unhampered by physical specialisation and honed by ice age hardship, they will become the most dangerous predators of all.

Gleaming here beneath a winter moon the snowfields of the Victorian Alps are now relatively small and wholly seasonal. During glacial winters snow would have covered most of the State for long periods.

BOGONG HIGH PLAINS, VIC.

FIRE & ICE

DUST STORM, STUART RANGE, S.A.

FIRE & ICE

5 MILLION YEARS TO ———— (2358 HOURS – ————)

LAND AREA DURING
LOWEST SEA LEVELS

(LEFT): When Australia began its solo voyage from the Antarctic, this was a region of forests and spring-fed streams. The few springs that remain now rise to a surface landscape that seems more suited to the moon.

Dust storms such as this (ABOVE), rather than winter's chill, best symbolise the current Ice Age in Australia. The denuded landscapes which yield the dust seem to have originated well before the growth of permanent snowfields in Australia.

(OVERLEAF): A degree of fire and drought resistance became a prerequisite for survival across much of inland Australia about five million years ago as the seasonal temperature range approached that of today. Hardy grasses, such as those commonly known as spinifex, had appeared, and eucalypts, casuarinas, and acacias were coping best with the weather changes.

WHERE ONCE THERE WERE shining lakes and broad, winding rivers, there are only mirages; willy willies play where rainforests once stood; and on barren hillsides, unshaded rocks crack open in the furnace of relentless desert. Australia is the most naked of continents. Five million years of desiccating westerlies have stripped away all but the hardiest of its vegetation. There is little trace of the well-watered continent that once lay so green and lush alongside Antarctica.

While satellite photographs of Australia's arid interior reveal faint traces of the extensive drainage patterns of old, survival in the open became impossible for the original vegetation. Some rainforest species found refuge along more humid coastlines or in sheltered valleys and gorges, but the safest escape route of all was upward. Climb the mountains of New Guinea or north-eastern Queensland and you climb into the past. Many old rainforest species, both plant and animal, kept ahead of the growing aridity by dispersing higher and higher up the slopes of the Great Dividing Range. Because they were more accustomed to cold, and to the poor soils of Gondwana, they tended to cluster on the highest peaks, forming 'island' populations, surrounded by a sea of better adapted modern forms. The mountainous coastline at the base of Cape York Peninsula is characterised by such beleaguered populations of old southern refugees.

Though the continent had been slowly drying out ever since it left Antarctica, by five million years ago there was still little true desert in Australia. Most of the interior was covered by drought-resistant vegetation that ranged from grassy scrub to open forest. Both grazing and browsing mammals thrived.

The trend towards gigantism in animals reached its peak at this time, the largest among them being the browsing marsupials. Some of

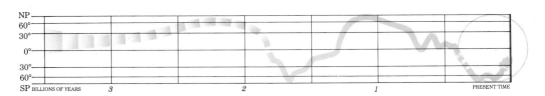

NP
60°
30°
0°
30°
60°
SP BILLIONS OF YEARS   3   2   1   PRESENT TIME

MARSUPIAL 'LEOPARD', *Thylacoleo carnifex* (CAST).

these animals, which ranged in size to that of a rhinoceros, may well have been equipped with short trunks and prehensile tongues, like the tapirs of South-east Asia and South America. Others were simply large versions of modern forms, both marsupial and monotreme. The platypus, for example, was very similar but several times the size of the modern species. It was only semi-aquatic, however, and had not yet lost its teeth.

Relatively few large predatory mammals roamed Australia at this time. There were two leopard-like marsupials, the *Wakaleo* and the *Thylacoleo* – the Tasmanian 'tiger' or Thylacine – as well as a meat-eating potoroid kangaroo, *Propleopus*.

The two largest carnivores, however, were both reptiles. One was a relative of modern pythons and boas, which grew about five metres long and up to 30 centimetres in diameter. The other, a member of the lizard family, was the largest four-legged land reptile since the dinosaurs. This predator would have dwarfed its largest modern relative, the huge Komodo Dragon, which survives on four small Indonesian islands. Only one species has been identified. Known as

SKULLS: *Diprotodon optatum* WITH *Sminthopsis sp.,* NAT. MUSEUM, VIC.

The dominant marsupial predators during the ice age appear to have been the Tasmanian 'Tiger' and this leopard-like tree-climbing hunter (LEFT), the Thylacoleo. While some doubt still surrounds its relationship to other marsupial groups, there is no doubt about the purpose and efficiency of its front fangs and the two large,

knife-edged molars with which it sheared the meat from its victims.

The tiny skull of the modern desert predator (BELOW) gives scale to the largest of all marsupials, a member of the Diprotodon family. This specimen was not yet fully grown when it died in a southern Victorian swamp several thousand years ago.

KANGAROO SKELETON, PILBARA, W.A.

The dividing line between survival and extinction was often fine indeed and hinged on factors which, in milder climates, may have been minor and ecologically meaningless. One of these shows up in several desert shrubs which, instead of drooping during their dry-season die-back, fold their branches upwards until they meet at the top (BELOW RIGHT). The tangle of branch tips at the apex then forms a natural shade house for the tender regrowth that springs from the root crown after the next good rains.

For those old Gondwanan species such as this Pandani (LEFT) that lived in the mountains, the ice age represented merely a return to 'normal': these were the kind of climates in which they had evolved when Australia lay with Antarctica beside the south pole.

Megalania prisca, it reached lengths of at least seven metres and weighed well over half a tonne. It was the main scavenger of Australia's arid regions.

As the ice-age deepened, animals had to cope not merely with regular water shortages but with widening extremes of temperature and frequent drought. Mammals especially had to be fast moving, fast breeding and opportunistic in their feeding. Neither predator nor herbivore could afford to overspend their energy reserves to gain a mere morsel, and many learned to go into a state of torpor during extreme climatic stress.

With a few exceptions, such as the Marsupial Mole, most of the adaptations to extreme aridity shown by Australian plants and animals are relatively superficial and probably evolved during the last five million years.

DESERT 'SHADEHOUSE', AMADEUS BASIN, N.T.

PANDANI, *Richea pandanifolia*, TAS.

THE OPPORTUNISTS

*O*pportunism in one form or another became the key to survival throughout most of the continental interior. The transformation of the landscape after rain is now one of the most memorable features of the Australian outback where floral carpets follow in the wake of major storms.

Exploding into life like the flowers, animal opportunists such as the Shield Shrimp populate the desert waterholes and claypan pools in a matter of days, (RIGHT) producing a seething hierarchy of ephemeral life. By the time the desert pools are dry again most have reproduced several generations: and a new crop of their dust-sized eggs lies embedded in the mud, awaiting desiccation.

EVERLASTINGS, *Helipterum sp,* AND TALL MULLA MULLA, W.A.

STURT'S DESERT PEA, *Clianthus formosus,* W.A.

WEEPING MULLA MULLA, *Ptilotus calostachyus,* W.A.

SHIELD SHRIMPS, *Triops australiensis,* W.A.

The roots of the fig family have made them one of the plant world's great survivors. By vigorous root growth they may find food at an inordinate distance from the trunk. In the case of the Strangler Fig, its aerial roots give it a head start in the struggle to reach light in the forest canopy. Many, like the Curtain Fig of north-eastern Queensland, eventually kill their original hosts, and must then depend on their roots for support.

SANDSTONE FIG, *Ficus platypoda*, W.A.

CURTAIN FIG, *Ficus virens*, QLD.

Displaying one of the most graphic adaptations to aridity by an Australian plant, the cactus-like appearance of this rare desert dweller thoroughly conceals its close relationship to the common hibiscus. It is known as Dunna Dunna, and grows only along the margins of a few claypans at the edge of Western Australia's Gibson Desert. In cutting its water budget to a minimum it has reduced both leaves and flowers to a dense layer of miniscule foliage that entirely sheathes the main stems.

DUNNA DUNNA, *Lawrencia helmsii*, W.A.

DUNNA DUNNA, *Lawrencia helmsii*, W.A.

The drying out of Australia has been a slow process and subject to many temporary reversals, but ultimately the transformation has been dramatic indeed. Where forest giants once lined permanently flowing creeks and rivers, dust-laden winds now sweep over a semi barren land.

The agents of change have not been merely a general reduction in the annual rainfall, or its increasingly episodic nature; in latter stages fire, floods, rising salt levels and desiccating winds have played a major role.

MONSOON STORM, TANAMI DESERT, W.A.

GYPSUM, NEAR LAKE EYRE, S.A.

As the land's internal drainage system dried out, salt levels rose and the remains of eastern Australia's huge inland seas were reduced to a series of brackish lakes. (The largest of these are Lake Eyre, Lake Gregory, Lake Blanche, Lake Callabonna and Lake Frome, all in South Australia.) The rising salt level increased the environmental stress on both plants and animals and a degree of tolerance to salt became a prerequisite for survival in many inland regions.

Another factor that filtered out Australian survivors was a need to withstand the fires that now ravaged huge tracts of land after electrical

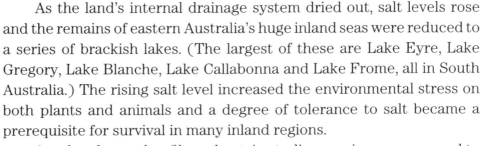

BURNT RAINFOREST, L. GORDON, TAS.

FORMER FOREST AREA, SIMPSON DESERT, S.A.

SAND DUNE, SIMPSON DESERT, S.A.

The growth of Australia's major dune systems probably began no more that two million years ago, although they now occupy vast areas of the interior as well as several coastal stretches.

In graphic testimony to the altered climate on Australia's west coast, weathered pillars of cemented sand are gaunt reminders of a forest that once grew here at a higher level. They represent vertical seepage channels that formed inside the sand dune, below the places where tree roots had broken through a semi-impervious surface crust. Carbonates dissolved from the upper layers then became re-deposited along the seepage channels, bonding the sand into these durable columns (RIGHT).

BEACH DUNES, EUCLA, W.A.

PINNACLES DESERT, NAMBUNG NATIONAL PARK, W.A.

PINNACLES DESERT, NAMBUNG NAT. PK., W.A.

FIRE & ICE

As Australia's internalised river system washed more and more mineral salts into the ancient drainage sumps of the interior, their evaporites built vast, salt encrusted playas, such as Lake Eyre (LEFT). As aridity increased the rivers themselves became choked with salt (BELOW). (RIGHT): Thriving despite a thick collar of exuded salt, this mangrove seedling is part of a colony that lives far from the sea beside a salt-choked billabong on the edge of Western Australia's Great Sandy Desert.

(OVERLEAF): One of the largest and most ancient of the internal drainage sumps is the Amadeus Basin in central Australia. At its heart lies a vast ribbon of gleaming salt known as Lake Amadeus, a small reminder of the sea corridor that once washed through this old central mobile belt.

'LOST RIVER', CANNING BASIN, W.A.

GREY MANGROVE, *Avicennia marina*, W.A.

FIRE & ICE

The first Tasmanian glaciers of the current ice age began to slide out of their mountain lairs some 750,000 years ago. They were to carve the highlands into their modern form. It would take only a relatively small drop in the mean annual temperature to return such areas as Mount Mawson (BELOW AND RIGHT) to the icy wasteland that it was a mere 18,000 years ago.

MOUNT MAWSON, MOUNT FIELD NATIONAL PARK, TAS.

SNOW TEXTURE, BOGONG HIGH PLAINS, VIC

(OVERLEAF): A *river of ice carved this narrow curving valley from the ancient quartzites of the Arthur Range in south-western Tasmania, during the last series of glaciations. In the foreground is Lake Mars, and in the background Lake Pedder and Lake Gordon.*

SPRING, MOUNT MAWSON, TAS.

storms. Most animals that were adapted to the aridity could escape because they were either highly mobile or burrowed. Vegetation, however, had to be able to regrow from its scorched remnants—often merely the root crown—or rely on the prolific production of well insulated seeds to create a new generation. Some inland trees evolved a smooth, white, bark that was so heat-reflective that the tree could survive grass fires without a single scorch mark.

Meanwhile, far to the south an icesheet was accumulating on Tasmania's central plateau. Repeated glaciation during later stages of

287

JADE POOL, KUBLA KHAN CAVE, TAS.

One of the world's most opulently decorated cave systems, the Kubla Khan, owes its existence to the inexorable seepage of meltwater from the ice-age snowfields of northern Tasmania. Its extensive system of crystal studded chambers began to form more than two million years ago within a massive outcrop of 450 million year old limestone. Almost three kilometres of chambers and passageways have now been mapped and these are believed to connect with another cave system close by that would add at least a kilometre to the total length.

the ice age would turn the western and southern margins of the plateau into the continent's only true mountain wilderness. The icesheet left its mark underground too, and two million years of seepage sculptured Tasmania's ancient limestones into a series of spectacularly ornamented caves. One of them, known as the Kubla Khan, has been rated among the world's finest.

There were at least four major cycles of cold, each broken into a series of glacial surges. During these surges billions of tonnes of the world's water became locked into polar icesheets and mountain snowfields: evaporation loss in the ocean went unreplaced and sea levels fell dramatically as a result, sometimes more than 200 metres below pre ice-age levels.

During these episodes the whole of Australia's continental shelf became exposed, considerably increasing its total land area. The Arafura Sea and Bass Strait both emptied, exposing land bridges that linked New Guinea, and then Tasmania, with the mainland. Similarly, the South-east Asian peninsula extended to include much of Indonesia.

SHAWL FORMATION, KUBLA KHAN CAVE, TAS.

290

KOALA, *Phascolarctos cinereus*, VIC

Australia's modern animals are, like ourselves, products of the current ice age. And like us, they reflect the considerable evolutionary fine tuning that has allowed them to cope with the drastically altered climates and escalating environmental stress of the last five million years. Kangaroos and wallabies, for example, have evolved one of the most sophisticated and yet flexible reproduction systems of any mammal; the wombat went underground and became much smaller; its tree-climbing relative, the koala, evolved a gut capable of coping with a diet of nothing but gum leaves.

COMMON WOMBAT, *Vombatus ursinus*, N.S.W.

BENNETT'S WALLABY, *Macropus rufogriseus*, TAS.

FIRE & ICE

LEAF MOTH, *Oenochroma vinaria*, N.S.W.

LEAF-TAILED GECKO, *Phyllurus platurus*, N.S.W.

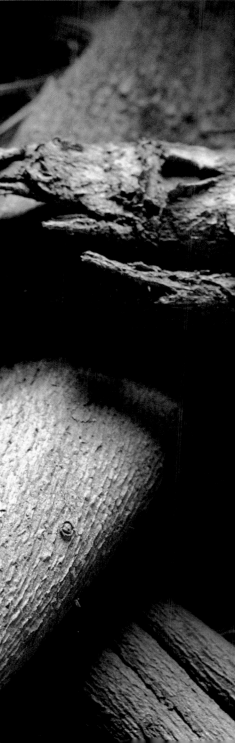

COATS OF MANY COLOURS

In response to the growing threat from the unstable climate and a procession of new, more dangerous predators, only those animals with the best defences survived. For the slow, the small and the defenceless, there was no substitute for good camouflage: the right combination of texture, colouring and body form could provide a passport into the future. Such deceptions still serve.

It would be hard to find better disguises than those used by the Leaf-tailed Gecko (BELOW), and the moth (LEFT). But there are some notable exceptions. The gaudy Leichhardt Grasshopper (RIGHT) of the Arnhem Plateau, advertises its presence with a coat of blue-black and crimson. Such a ploy signals to potential predators that this intended victim is either highly distasteful or even poisonous, and should be left alone.

LEICHHARDT GRASSHOPPER, *Petasida ephippigira,*

SPINIFEX HOPPING-MOUSE, *Notomys alexis*, N.T.

With characteristic efficiency the rodents have colonised almost every available habitat in Australia from desert to snow-fields, evolving forms that are very similar in some cases to those of their marsupial counterparts. The delicate hopping mice (BELOW) of arid Australia already resemble the marsupial Kultarr in some aspects, although the Kultarr is fiercely predatory while the mice are mainly vegetarian. Both have evolved a long tufted tail to help maintain their balance while they are airborne: however in this case it is the mice that hop and the marsupial that uses all four feet.

By contrast, the Broad-toothed Rat (BELOW), can lead a fully active life beneath a blanket of snow. This plump, long-haired species is now thoroughly adapted to a semi alpine existence in cold rainforest gullies and alpine heath.

BROAD-TOOTHED RAT, *Mastacomys fuscus*, TAS.

The interchange of land life between these two regions increased sharply during such periods. Some Asian plants and animals even reached Tasmania by these land bridges, Tasmania's rodent population being a typical example.

Rodents disperse fast and adapt quickly, talents that have made them one of the world's most successful mammals. The oldest sign of their presence in Australia appears in a five million-year-old fossil deposit known as Rackham's Roost, in north-western Queensland. These new invaders soon colonised most of the Australian continent, penetrating desert, rainforest and alpine heathland with remarkable efficiency. It was one of the swiftest conquests of all.

There would be only one swifter. On the drying plains of north-eastern Africa, evolution was tinkering once more with the timing mechanisms that controlled the juvenile development of a group of hominoid primates. About 2 million years ago, a new genus, *Homo*, appeared. Their descendants would not only dominate the entire planet, they would begin to explore the universe.

2 million years – present

AS THE FINAL SECONDS *of the time scale tick away, tides of terrible cold sweep repeatedly over the world's temperate regions. Stripped of fur, claws and fighting teeth, our hominid ancestors enter Europe ill-equipped to survive either the northern climate or the awesome predatory competition. Yet among the mixed bag of genetic 'deformities' that produced them is a solitary weapon, a curiously distorted primate brain. With crude technology, a little cunning and a crucial touch of madness, they tackle mammoth and sabre-toothed tiger with remarkable success. Slowly they spread to the farthest ends of Asia. For a time this is the limit of their advance but as the last and most savage glaciation begins, it is clear that not one, but two very different kinds of human beings have arrived in Australia. One is from Java; the other seems to be of modern Chinese origin. Both survive the cold and final thaw: it is the drying out that proves to be the hazard. The campfires of the Javan hominids eventually flicker and die, a people lost forever along with the last of the old marsupial giants. The Aborigines, a blend of many invaders, prevail on this driest of continents by becoming the world's most sophisticated hunter-gatherers.*

Some concept of life after death has characterised human societies for at least 60,000 years. The Ngarrenjeri people of South Australia preferred to face their afterlife sitting upright with knees drawn up against the chest.

ABORIGINAL BURIAL GROUND, COORONG, S.A. (PHOTO: ROB MORRISON)

BEHOLD, THIS DREAMER . . .

GENESIS 37, VERSES 19, 20

SHELL MIDDEN, LAKE MUNGO, N.S.W.

BEHOLD, THIS DREAMER...

HUMAN ENTRY ROUTES
FOSSIL SITES

This granite boulder (LEFT) and the elegant design etched into its surface represents a span of 3.5 billion years. The granite crystallised beneath the Pilbara as life first began to flourish along its shores, while the design was the brainchild of one of life's latest species.

Just as the granite coincided with the dawn of biological evolution, so the etching of designs such as this signalled the advent of a new kind of evolution, one based not on the transmission of genes from one generation to the next, but on the transmission of ideas.

The remains of shellfish meals (ABOVE), fragments of campfire charcoal, stone tools and two human burial sites, have confirmed Lake Mungo as one of the oldest known campsites of fully modern human beings in the world.

(OVERLEAF): The eroded dunes known as the China Walls started to blow away as the snow-fed lakes that they fringed began to dry up some 12,000 to 15,000 years ago.

AS YOU CLIMB the crusty slope that rims the eastern edge of the claypan, you enter a landscape that seems more suited to the moon than to south-western New South Wales. The big sand dunes that once dominated here have in recent times been stripped by incessant westerlies, leaving only a scattering of gaunt pinnacles. Little grows and almost nothing moves, except the sand as it smokes along the ground and coils about your legs in prickling gusts. But if you could peel away the years to match some of the older sand horizons currently exposed, you would find yourself on the edge of a lake. Thirty thousand years ago trees and rushes lined its shores and clouds of water birds fed in the shallows. Moreover, among the dunes along its eastern side, the wind often carried wisps of wood smoke and the unmistakable smells of the cooking fire: for this is Lake Mungo, one of the oldest known campsites of modern human beings anywhere in the world.

The 4-billion-year trail that leads from the Jack Hills and Mount Narryer zircon crystals to the decaying dunes of Lake Mungo has many milestones. All are memorable, but few are of greater significance for our species than the skeleton of a young man, buried with some ceremony beside the lake more than 30,000 years ago.

He was light boned and gracile, with the kind of domed forehead and flattish face that would pass unnoticed in any modern crowd. The only older skulls that show comparable development have come from cave deposits near the Li River in southern China. They are, in fact, so like the Mungo skulls as to suggest a close genetic link.

Conflicting theories abound, but most experts believe that modern *Homo sapiens* arrived in South-East Asia between 50,000 and 60,000 years ago. Their appearance in southern Australia so soon afterwards suggests that they were imaginative and resourceful, built efficient sea-going rafts,

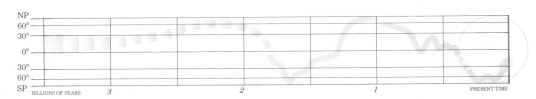

ABORIGINAL PETROGLYPH, PILBARA, W.A.

and possessed complex language. In other words, Mungo people were like us, clearly classifiable as *Homo sapiens*. Yet just 270 kilometres south-east of Lake Mungo, beside the Murray River, skulls unearthed on the fringes of a drowned forest known as Kow Swamp reveal a very different kind of human being. The architecture of these skulls suggests that their owners belonged to a relatively archaic branch of the human family. One that had long since died out elsewhere in the world.

It has been proposed that the distinction between the Lake Mungo and Kow Swamp skulls demonstrates a greater evolutionary gulf than that which separates us from the Neanderthals of ice-age Europe. The inescapable corollary is that for many thousands of years Australia was inhabited not by two different races, but by two different branches of the human family tree.

Kow Swamp people were short, stocky and heavy browed, with sloping foreheads and robust but receding jaws. They most resemble hominids that lived near the Solo River in central Java about 300,000 years ago. Micro-evolution within that relatively isolated population of Javan archaics appears to have imparted a distinctively regional character to their descendants, and it is this character which reappears in modified form at Kow Swamp. These people moved into the area about 13,000 years ago, and signs of their occupation cut out again 4,000 years later. Skulls bearing the same stamp of Javan ancestry have since been found at other sites, one near Cossack on Australia's north-west coast, and one in southern Queensland. Yet curiously, most of these relics are roughly contemporary. So it is the very large time gap between the disappearance of the Javan hominids and the occupation of Australia by their apparent descendants that is most tantalising.

Reliable evidence of humans in Australia is currently no older than 43,000 years, although a small piece of hand-ground haematite found in

(LEFT) The skull in front was found at Kow Swamp beside the Murray River: behind it is a skull from Keilor, near Melbourne. There is greater structural difference and more evolutionary distance between them than there is between ourselves and the Neanderthals that lived in central Europe 100,000 years ago. Both skulls are about 13,000 years old.

This drowned forest provided a rich hunting ground as recently as 9,000 years ago for a strain of human being that had died out elsewhere in the world.

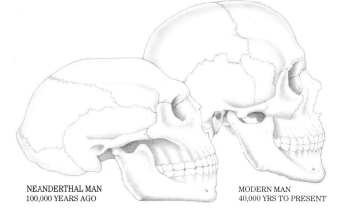

NEANDERTHAL MAN
100,000 YEARS AGO

MODERN MAN
40,000 YRS TO PRESENT

Arnhemland has been tentatively dated at almost 60,000 years old. However, this figure was achieved by means of a highly problematic technique known as luminescence dating. There are two forms of it, and both measure the degree of luminescence that grains of sand gradually acquire when they become reburied after their electron 'clocks' have been reset by exposure to heat or light.

There is, however, one other major clue. Some lake and marine sediments show a distinct horizon made up of two factors: the amount of burnt matter—carbon—that the sample contains, and the proportion of

The generations of ice age hunters that took shelter in this cave sharpened their stone weapons on rock walls of glacial debris that had been left by the previous ice age, some 250 million years earlier.

ABORIGINAL GRINDING GROOVES, KIMBERLEYS, W.A.

PAWN OF THE RED QUEEN

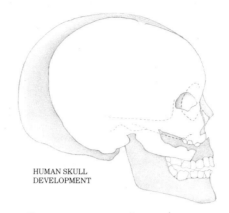

HUMAN SKULL
DEVELOPMENT

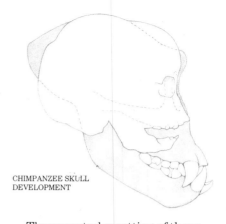

CHIMPANZEE SKULL
DEVELOPMENT

In Lewis Carroll's 'Through the Looking Glass', the Red Queen is forced to keep running in order to stay in the same place. On the vaster stage of the geological time scale, life faces the same problem. The environment is in a constant state of flux due to Earth's crustal movement, changing weather patterns and several extraterrestrial factors, such as the impact of comets and fluctuations in solar energy. Evolution is life's method of 'running' to keep up.

One of the keys to life's success has been the flexibility of its driving mechanism, the DNA code. This too changes constantly in all species. Small molecular variations appear to accumulate with time.

The form, colour and character of animals and plants are dictated by segments of their DNA code, known as genes. These act in committee groups, regulating all growth and bodily functions. However, while variation within these gene committees is responsible for variation between individuals, the kind of structural changes that lead to new life forms seem to arise from minor adjustments to the myriad chemical time switches that control the genes. These tiny molecular triggers, embedded at each end of a gene segment, are responsible for starting or stopping protein production within each cell.

The repeated resetting of these body clocks within our primate ancestors eventually produced our current human species. Clearest evidence of this lies in a comparison between the skull growth in a modern human being and that of a chimpanzee, our nearest living relative. The similarity between a baby chimp and a human baby is striking during the foetal stage. This similarity fades rapidly as they mature, with the chimp skull changing most. Despite a much longer growth period, the slow-growing human skull retains much of the character that is common to both babies. It is a classic case of neoteny, an evolutionary phenomenon in which sexual maturity overtakes a relatively juvenile body. Other 'juvenile ape' characteristics appear in the slender proportions of our bodies, under-developed canine teeth, ventral vaginas and finer body hair.

In this sense then we are 'juvenile apes', because analysis of some 40 blood proteins in both chimps and humans shows a difference of less than one per cent. According to rates of change that are common among primates, this minute variation suggests that it was not more than eight million years ago, and perhaps as little as six million years ago, when the Red Queen of evolution led with her new pawn, the hominid.

STONE ARTEFACTS, AMADEUS BASIN, N.T.

PUKUMANI POLES, BATHURST ISLAND, N.T.

Crucial to the hunter's arsenal were his fire sticks. Though the principle remained the same, the technique of fire-making, and consequently of fire-stick design, varied considerably in different parts of the continent. These examples are of a design used by the people of Cape York Peninsula in north-eastern Queensland. The fire-stick holder is decorated by seeds pressed into fire-softened resin (LEFT).

The Aborigines' stone tools and weapons had to be continually replaced, so regular campsites eventually became littered with broken grind stones, cutting tools and stone chips (ABOVE). These examples were found at a vantage point among sandhills, where hunters worked on their weapons while waiting to ambush game approaching a nearby waterhole.

These carved posts mark the site of a mortuary ritual performed at the death of a Bathurst Island leader of the Tiwi people (ABOVE RIGHT).

fire-resistant plant species represented in its fossilised pollens. Both levels rise fairly sharply between 120,000 and 110,000 years ago in Australian soils, confirming that fire became common around that time.

This sedimentary horizon coincides with a warm but relatively humid interglacial period and is unlikely to reflect a climate change. It seems much more likely that the increased carbon signals the arrival of hunters carrying fire sticks, and the transition between pollen types shows their dramatic impact on the vegetation.

A similar carbon horizon has since shown up in a drill core taken from Australia's continental shelf east of Cairns. It has yielded dates that lie between 130,000 and 140,000 years ago. Meanwhile, an even older carbon horizon has recently been identified in a drill core pulled from the seabed off the Indonesian island of Lombok. This showed a securely dated carbon peak about 200,000 years ago that was closely followed in the pollen record by an abrupt switch from rainforest tree species to grassland species.

Fire has served as a human tool for a least 350,000 years, and its regular use probably dates back 700,000 years. Meanwhile the occasional capture and opportunistic use of fire appears to extend back some 2 million years or more. There is therefore no reason to doubt that archaic Australians would have quickly grasped its peculiar potential here. Not only was it useful as a hunting tool to flush out game, but it offered them a 'farming' tool, since fire promoted seeding, germination and fresh growth in much Australian vegetation.

In fact, the only satisfactory explanation for these well-defined carbon-pollen horizons and their sequential southward march is that they signal the gradual advance of archaic hunter–gatherers armed with fire-sticks. The fire-farming techniques that they evolved here were copied by modern human beings when they finally arrived tens of thousands of years later. To deny that those archaic Australians were capable of using fire in this sophisticated fashion would seem to grossly underestimate them—especially in view of their migration record.

FIRE STICKS AND HOLDER, CAPE YORK PEN., QLD

PETROGLYPHS, EWANINGA, N.T.

MALE AND FEMALE FIGURES, N'DHALA GORGE, N.T.

The rock art of the Aboriginal people of Australia has accumulated over an enormous period, perhaps as much as 30,000 years. No paintings are likely to have endured for this long but some rock etchings are of great age. However, dating these is virtually impossible in most cases, though their style often gives a clue to when the artist lived and may even suggest what the environment was like at that time.

Many of the deeply-carved petroglyphs (ABOVE LEFT) that pattern the sandstone rock slabs, beside a desert claypan south of Alice Springs, are strongly representational and clearly recent. By contrast, the human figures pecked into the durable rock faces of N'Dhala Gorge in the eastern Macdonnell Ranges are both stylised and dominated by very large and ornate head decorations (BELOW LEFT), and appear to be many thousands of years old.

Most recent of all are the sophisticated X-ray style paintings that decorate many of the rock shelters frequented by the nomadic peoples of the Arnhem plateau. They are not merely representational in outline: they usually show something of the subject's internal structure (RIGHT).

X-RAY ART, UBIRR, KAKADU, N.T.

EARLY HUMAN ENTRY ROUTES

HOSKYN ISLAND, BUNKER GROUP, QLD

The gradual rise in sea level that accompanied the melting of the polar ice caps began around 16,000 years ago and lasted some 10,000 years. Though episodic in its early stages, it resulted in large-scale flooding in north-eastern Australia, allowing corals to re-establish themselves as the land submerged. They then kept pace with the rise of the water to build the most extensive reef system in the world. Like the rain-forests that fringe the coastline, the Great Barrier Reef now represents a storehouse of species and a major testing ground of evolution. However, the gardens of living coral (RIGHT) that decorate its surface represent only the tip of this massive structure, which reaches a depth of more than 200 metres in some places, and extends for a distance of 2,300 kilometres along Australia's north-eastern coastline.

Another link in the chain of evidence that points to Java as the source of Australia's first settlers was unearthed recently on the island of Flores, which lies some 700 kilometres east of Java. Archaeologists found more than a dozen hand-worked stone tools lying in layered sediments near the bones of a pygmy elephant. This find was conveniently sandwiched between two deposits of volcanic ash that have since been securely dated as 800,000 and 880,000 years old. Meanwhile Flores is part of the Wallacea chain of volcanic islands and has never been attached to mainland Asia. Humans would have had to cross three sea barriers, all of them more than 19 kilometres wide. With such clear evidence of migration by sea in Australia's direction at least 800,000 years ago, it seems unlikely that descendants of those adventurous seafarers would have failed to reach Australia in the next 700,000 years.

Besides, if fully modern human beings had arrived first, the archaics could not have achieved the wide distribution that their fossils now display. Modern humans would have viewed the archaics as alien, inferior creatures and would have dealt with their invasion in traditional *Homo sapiens* fashion. Aggressive territoriality is what works best for meat-eating species, and having a significant mental edge on their archaic opponents, the home team could not have lost. But if the archaics were indeed the original occupants, and were invaded by modern humans, they would have retreated to the inhospitable heartlands, just as the Aborigines did when Europeans arrived some 40,000 years later.

SOFT CORAL, LIZARD ISLAND, QLD

SEA CUCUMBER, *Pseudocolochirus axiologus*, QLD.

Coral and marine life become
more diverse the further offshore
and the further north that they are
found. Reefs near Lizard Island
off Cooktown boast a variety of
corals and other marine life that
equals those of most South Pacific
reefs. One of the reef's most
colourful inhabitants (ABOVE) is
this sea cucumber. It has only one
body opening amid its forest of
feeding arms. Through this it both
'inhales' its food and 'exhales' its
waste products. Its relatives
include the trepang, much prized
by the Macassan fishermen, who
searched Australia's northern
shores for it long before the first
European settlers arrived.

FRINGE REEF, EAGLE CAY, QLD.

McLARTY RA., KIMBERLEYS, W.A.

The rising seas that marked the end of the last glaciation inundated almost one seventh of the continental raft, completely redefining the outline of Australia. The Kimberley archipelagos are a good illustration of this (RIGHT).
They are the tops of hills that once lay far from the sea, and their silt-filled valleys are now etched with drainage patterns left by the huge tides that wash this coast (BELOW RIGHT). Most dramatic of all, however, are the two drowned river gorges that cut through the ridge lines of the McLarty Range north of Derby (BELOW). The sea thunders in and out of these gorges twice a day as the tide fills and empties the two drowned valleys that they link (LEFT).

TIDAL RAPIDS, McLARTY RA., W.A.

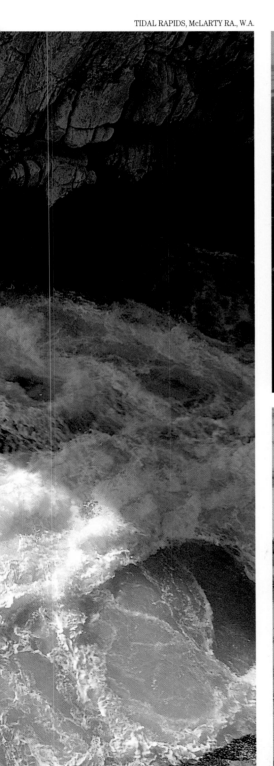

BUCCANEER ARCHIPELAGO, W.A.

TIDAL DRAINAGE, KIMBERLEYS, W.A.

OCHRE MINE, MACDONNELL RANGE, N.T.

Bonded to the land by their complex totemic culture which embraced the landscape itself, the Aboriginal evolved a strong bond with the materials that formed it. Ochre represents much more than a vehicle for adornment or for symbolic representation. The mining, grinding and application of it assumes a ritualistic importance that, in some cases, makes it an intrinsic part of ceremonial events (BELOW LEFT) and (RIGHT).

Good quality ochre such as that from the traditional mine in the western Macdonnell Ranges was highly prized all over Australia and consequently became a valuable trading commodity between tribal groups (LEFT).

APPLYING OCHRE, KINTORE, N.T.

LEAD SINGER, WOMEN'S CEREMONY, KINTORE, N.T.

We may never discover what ultimately became of Australia's archaic inhabitants, but the fact that they disappear from the fossil record 9,000 years ago may well be significant. This was a time of rising sea levels and desiccation in Australia's interior. Almost one sixth of the continent became inundated, and most of this lost area would have been prime coastal land occupied by *Homo sapiens*. Coastal clans would have continually edged inland as seas rose, further increasing the pressure on those few archaics that had managed to survive the climate change.

Proof of at least one further *Homo sapiens* invasion from the north is provided by the arrival of the dingo on mainland Australia between 4,000 and 5,000 years ago. This in turn coincided with the decline of the two dominant marsupial carnivores, the Thylacine and the Tasmanian Devil. However, protected by the flooding of Bass Strait, the Tasmanian populations of these two species remained secure. Neither the new human migrants nor their dogs seem to have made the crossing.

Meanwhile, on the mainland, so effectively did this composite of modern human colonists interweave their mythology, social structures and hunting techniques, that their descendants, the Aborigines, became the most sophisticated hunter–gatherers the world has produced. Their long apprenticeship in opportunistic survival nurtured remarkable powers of observation, memory and deduction in matters related to their natural environment. And it built a system of family and tribal relationships so intricate and binding that it remains beyond the grasp of most outsiders. It also served to bond them to the land in a totemic relationship that inverted all the rules of ownership that characterise 'civilised' societies. Under this unique contract, the land owned them.

The price of such extreme specialisation became apparent on 18 January 1788, when a small wooden sailing ship, the first of many, entered Botany Bay.

(LEFT): The body decoration worn by this man of the Tiwi people on Bathurst Island, is as unique and personalised as a fingerprint, although the use of individual designs is strictly governed by the wearer's totemic rights and by the occasion itself.

This representation (BELOW) of a Macassan prau and its tender may well be more than 200 years old. The Macassans were the first regular visitors to Australian shores, attracted by the good fishing and the abundance of trepang, or sea cucumbers, which they prized as a delicacy. But their bleached cloth sails were an omen of another kind of exploration that was just beginning.

NANGALOAR SHELTER, KAKADU, N.T.

Epilogue

I HAVE EXCHANGED the soughing of the night wind in desert oaks for the dull roar of suburbia, the howl of the territorial dingo for the wail of the ambulance. My journey is over, and the spell of my years on the road has faded. My study window faces south, but where southern stars should trace their fiery arcs in the velvet night, there is little now to remind me of the real world that lies beyond these polluted skies. But one image does come to mind – an eroded, rust-red sea bed that lies 4,000 kilometres to the west. On most maps the only reference to it are the words 'Hamersley Range'. It is their evolutionary significance that haunts me. This ancient seabed commemorates life's first explosive success on this fertile but finite planet. Life would pay dearly for that success. It was a time when vast blooms of photosynthetic bacteria overloaded the primitive seas with their oxygen waste. The oxygen reacted with the soluble iron, triggering a blizzard of marine rust that lasted half a billion years. Poisoned by their own waste gas, many early forms vanished from the fossil record. Nevertheless, life recovered and evolution continued, but its course had been changed forever. Curiously, that biological turning point is commemorated in every cell of our bodies, for we are one of its by-products.

And other visions drift in through my study window. A virgin rainforest clear-felled, waiting for the wood-chippers; a mountain stream gurgling through rusty cans and car bodies; clear artesian waters bleeding endlessly to satisfy herds of destructive, hard-hooved stock; a chain of toxic waterholes where Australia's greatest river used to run until the irrigators arrived, and everywhere the rust-brown wind foams across a denuded land. For every evolutionary success there is a fee, and we are facing the fee for ours.

The wheels of biospheric retribution grind very slowly, and we may already have passed the point of no return. But if our species is to survive, our last chance surely lies in assigning a realistic value to the natural environment and preserving the life that's left on this beleaguered planet.

STROMATOLITES, SHARK BAY, W.A.

SANDSTONE LAYERS, MURCHISON GORGE, W.A.

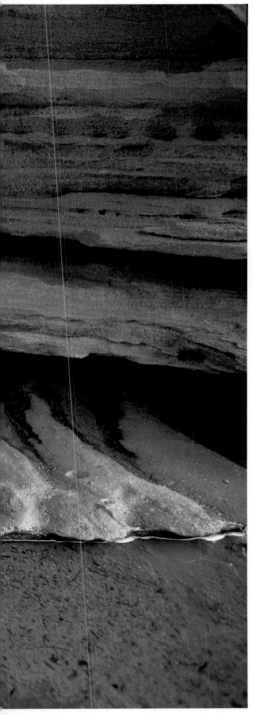

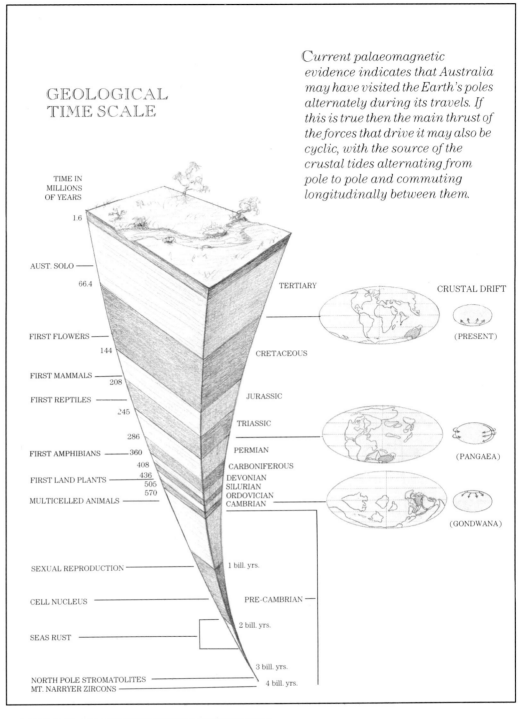

GEOLOGICAL TIME SCALE

Current palaeomagnetic evidence indicates that Australia may have visited the Earth's poles alternately during its travels. If this is true then the main thrust of the forces that drive it may also be cyclic, with the source of the crustal tides alternating from pole to pole and commuting longitudinally between them.

TIME IN MILLIONS OF YEARS

1.6

AUST. SOLO — 66.4

TERTIARY

CRUSTAL DRIFT

(PRESENT)

FIRST FLOWERS — 144

CRETACEOUS

FIRST MAMMALS — 208

JURASSIC

FIRST REPTILES — 245

TRIASSIC

286

PERMIAN

(PANGAEA)

FIRST AMPHIBIANS — 360

CARBONIFEROUS

408

DEVONIAN

FIRST LAND PLANTS — 436

SILURIAN

505

ORDOVICIAN

570

CAMBRIAN

MULTICELLED ANIMALS —

(GONDWANA)

SEXUAL REPRODUCTION — 1 bill. yrs.

CELL NUCLEUS — PRE-CAMBRIAN

2 bill. yrs.

SEAS RUST —

3 bill. yrs.

NORTH POLE STROMATOLITES — 4 bill. yrs.
MT. NARRYER ZIRCONS —

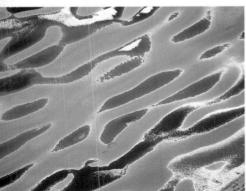

TIDAL SAND BARS, BRUNY IS., TAS.

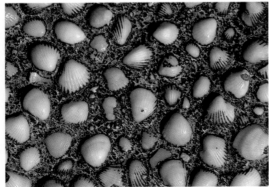

ALGAL MAT, SHARK BAY, W.A.

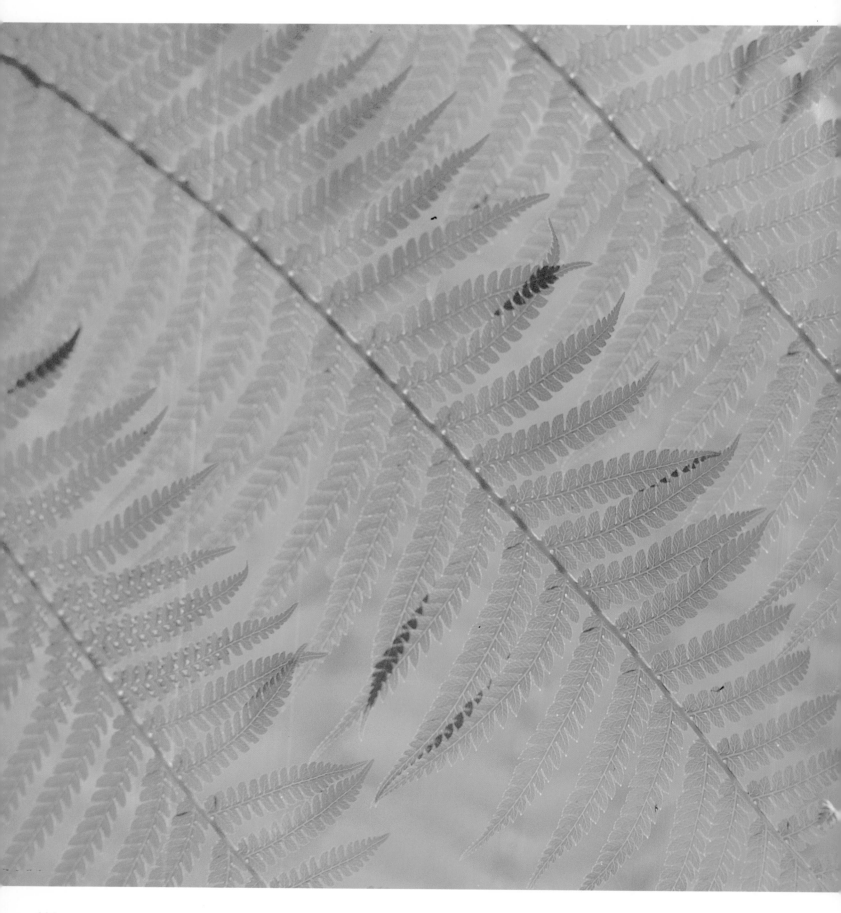

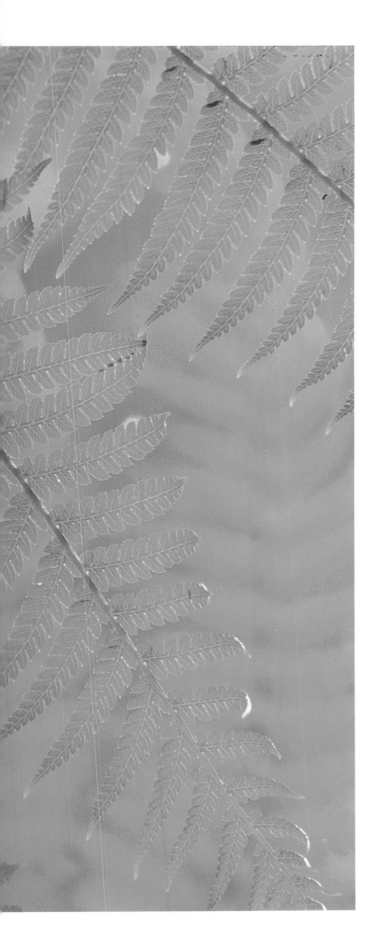

BIBLIOGRAPHY

ARCHER, Michael & CLAYTON, Georgina (eds.) Vertibrate Zoogeography and Evolution in
    Australasia. Hesperian, Perth, 1984.
ATTENBOROUGH, David. Life on Earth. Little, Brown & Company, Boston, 1983.
BERNDT, R. M. & C. H. The World of the First Australians. Lansdowne, Sydney, 1981.
BLAINEY, Geoffrey. Triumph of the Nomads. Overlook Press, New York, 1982.
BRONOWSKI, J. The Ascent of Man. BBC, London, 1973.
BROWN, CAMPBELL & CROOK. Geological Evolution of Aust. & NZ. Pergamon, Sydney, 1969.
CLARK, I. F. & COOK, B. J. (eds.). Perspectives of the Earth. Aust. Academy of Science,
    Canberra, 1983.
CROWELL, J. C. & FRAKES, L. A. Late Palaeozoic Glaciation (paper).
DAWKINS, Richard. The Selfish Gene. Oxford University Press, London, 1976.
FRAKES, L. A. Climates Throughout Geological Time. Elsevier, Amsterdam, 1979.
GRASSWILL, Helen. Aust. Timeless Grandeur. Lansdowne, Sydney, 1981.
HOYLE, Fred. WICKRAMASINGHE. Chandra., Life Cloud. J. M. Dent, London, 1972.
HALLAM, J. P. (ed.), Planet Earth. Rigby, Adelaide, 1977.
ISAACS, J. (ed.), 40,000 Years of Australian Dreaming. Lansdowne, Sydney, 1980.
KEAST, Allen. (ed.). Ecological Biogeography of Aust. Dr. W. Junk bv., The Hague, 1981.
KRONER, A. (ed.). Precambrian Plate Tectonics. Elsevier, Amsterdam, 1981.
LASERON, Chas. Ancient Australia. Angus & Robertson, Sydney, 1969.
LEAKEY, Richard, LEWIN, Roger. Origins. E. P. Dutton, New York, 1982.
MACDONALD, M. R. The Origin of Johnny. Jonathan Cape, London 1975.
MORLEY, B. D., TOELKEN, H. R. (eds.) Flowering Plants in Australia. Rigby, Adelaide, 1983.
PONNAMPERUMA, Cyril. Origins of Life. Thames & Hudson, London, 1972.
PLUMB, K. A., Bureau Min. Rec. Canberra. The Tectonic Evolution of Australia (paper).
    25th International Geol. Congress, Sydney, 1976.
QUIRK, Susan, ARCHER, Michael, SCHOUTEN, Peter (eds.). Prehistoric Animals of
    Australia. Australian Museum, Sydney, 1983.
SAGAN, Carl. Cosmos. Random House, Inc., New York, 1983.
SCHREIBNER, E. (ed.). Tectonophysics: The Phanerozoic Structure of Australia.
    Elsevier, Amsterdam, 1978.
STRAHAN, Ronald (ed.). Complete Book of Australian Mammals. A&R, Sydney, 1983.
TAKEUCHI, H., EDEYA, S., KANAMORI, H., (eds.). Debate about the Earth. Freeman
    Cooper, San Francisco, 1970.
TARLING, Dr. D. H., Continental Drift. Redwood, London, 1971.
THOMAS, Barry. Evolution of Plants and Flowers. Eurobook, Madrid, 1981.
THOMAS, R. D. K. When and how did plants and animals take to the land? (paper).
    Paleontological Soc., Lancaster, Penn., 1984.
VEEVERS, J. J. (ed.) Phanerozoic Earth History of Australia. Clarendon, Oxford, 1984.
WALTER, M. R. (ed.). Stromatolites. Elsevier, Amsterdam, 1976.
WATSON, J. D., TOOZE, J., KUTZ, D. T. Recombinant DNA, a short course. Scientific
    American, New York, 1983.
WHITE, Mary E. Australia's Prehistoric Plants. Methuen Aust., Sydney, 1984.
    Greening of Gondwana. Reeds Sydney, 1986.
WILD, J. P. (ed.) In the Beginning. Aust. Academy of Science, Canberra, 1974.

PALM FROND SHEATH, QLD.

INDEX

Illustrations are indicated in italics